Indigenous Knowledge Systems, Entrepreneurship, and Sustainable Development in Africa

Samuel Ojo Oloruntoba, Inocent Moyo,
Zainab Monisola Olaitan and Jabulile H Mzimela

ANTHEM PRESS

Anthem Press
An imprint of Wimbledon Publishing Company
www.anthempress.com

This edition first published in UK and USA 2026
by ANTHEM PRESS
75–76 Blackfriars Road, London SE1 8HA, UK
or PO Box 9779, London SW19 7ZG, UK
and
244 Madison Ave #116, New York, NY 10016, USA

British Library Cataloguing-in-Publication Data
A catalogue record for this book is available from the British Library.

Library of Congress Cataloging-in-Publication Data: 2026941948

ISBN-13: 978-1-83999-681-8 (Hbk)
ISBN-10: 1-83999-681-1 (Hbk)

This title is also available as an eBook.

CONTENTS

FOREWORD

This book complements existing scholarship on the increasing recognition of indigenous knowledge in addressing most of the inscrutable challenges of our time, such as poverty, unemployment, inequality, and climate change. It responds to the limited success of applying exogenous knowledge largely rooted in Eurocentrism in tackling socioeconomic challenges in Africa. Despite centuries of epistemic othering and subordination to which indigenous knowledge in Africa was subjected, it has continued to show resilience, relevance, and resourcefulness. Indigenous knowledge in Africa remains relevant not only because it is rooted in the sociocultural and historical experiences of the people, but more importantly because it provides an alternative path of survival for the majority that are excluded from the neoliberal capitalist economic order that subsists from colonial to contemporary times.

While this neoliberal order has created opportunities for a few who are educated and occupy positions in public and private sectors, there is a vast majority of Africans who exist at the margins. Notwithstanding the coloniality of knowledge and power that subsists within the postcolonial state structure, there are incipient, if quiet, decolonial efforts by Africans at the margin through which they continue to engage with and apply their indigenous knowledge to meeting basic needs such as economic livelihoods, trading, raising funds for businesses, managing the environment, and health. The unending entrapment of the state in coloniality through unceasing entanglement with policies flowing from the neoliberal economic order and its disempowering proclivities necessitates decolonization of knowledge on economic practices, natural resources governance, climate policy, and business financing.

This book highlights the imperative of decolonization in the ways in which economic activities such as entrepreneurship, the role of women, and how climate change affects livelihoods are framed in Africa. It underscores the essence of co-creation in the design of policy and the intentional infusion of indigenous knowledge on how the state supports these activities. While the focus of the state in Africa has been to promote entrepreneurship based on neoclassical economic theory of profit maximization and reification of individual achievement, this book expatriates on indigenous entrepreneurship where profit-making coexists with communal sharing, protection of the environment, and consideration of the future.

Additionally, the book posits that the overall failure of development practice in Africa is rooted in its extraversion and the ontological framing in neoliberal capitalist philosophy of self-accumulation and the pursuit of materiality for the sake of it. Whereas development has been one of the most popular political and economic discourses in Africa, it has not been successful because it was not rooted in Africa's knowledge system, nor has it been co-designed in partnership with the people. Thus, the book reconstructs development away from merely growth in gross domestic product and other popular indices to include recognition of the importance of co-creation, voice, and agency of the people, culture, language, and spatial inclusion. In other words, while incremental change in human conditions through advancement in technology, economic productivity, exchange, and reliable institutions is important, their value is limited in the contexts in which the knowledge that forms the foundation of a society is neglected. Decolonizing discourses and practices of sustainable livelihoods is critical to epistemic freedom, inclusive transformation across various categories such as gender, youth, urban, and rural, as well as policy. The authors bring to bear years of policy and scholarly engagements in weaving together the various ways in which indigenous knowledge can foster indigenous entrepreneurship and overall sustainable entrepreneurship in Africa.

ACKNOWLEDGMENTS

This book flows from the project on Unlocking Sustainable Solutions for Employment and Entrepreneurship: Empowering African Youth Through Indigenous Knowledge, managed by the Africa Indigenous Knowledge Research Network and supported by the Mastercard Foundation. The views expressed do not necessarily represent those of the Foundation, its staff, or its Board of Directors. The authors acknowledge the insights from the various communities where the research took place, as well as the various perspectives shared by members of the network during our webinars and conference.

Chapter 1

INDIGENOUS KNOWLEDGE SYSTEMS, ENTREPRENEURSHIP, AND SOCIOECONOMIC DEVELOPMENT IN AFRICA: CONTENTIONS AND DEBATES

Introduction

In a book whose focus is on Indigenous Knowledge Systems (IKS), entrepreneurship, and socioeconomic development in Africa, a good starting point is a definition of what IKS and Indigenous entrepreneurship are. While it is accepted that the notion of IKS is a matter that has generated and continues to generate debate (see, e.g., Mawere 2015; Mwangi, 1998), nonetheless, it is regarded as a "body of knowledge built through generations living in close contact with nature and it includes a system of classification, a set of empirical observations about the local environment, and a system of self-management that governs resource use" (Mwangi 1998:2). IKS is characterized by its inherent integration into people's experiences within a specific geographic location and its transmission through oral tradition between generations, which allows it to evolve and/or improve over time (Mwangi 1998). As such, IKS is different in various locations. The corollary is that societies in Africa have different IKS than those in Australia, Asia, or Latin America inter alia. More so, within Africa, Latin America, and Asia, IKS varies from one Indigenous community to the other.

Another way to look at this is that IKS has a contextual and geographical specificity. Hence, the correctness of the observation by Senanayake (2006:87), that IKS "is rooted to a particular place and set of experiences and generated by people living in those places. The result of this is that transferring the Indigenous knowledge to other places runs the risk of dislocating it." IKS also has the characteristic of evolving and developing depending on real-world experiences and application (see, e.g., Mawere 2015; Mwangi, 1998). This also means that IKS continues to be used in different contexts and locales (see, e.g., Moyo, 2022). This is more so the case because IKS is employed in the use and/or governance of resources as well as in ethics of self-care (Mwangi 1998:2). From this definition, it is evident that IKS is a body of knowledge that people can use in transacting in this life including building communities and the generation of livelihood opportunities (see, e.g., Dana, 2015; Mawere 2015; Mwangi, 1998)—we will return to this point.

Indigenous entrepreneurship is the generation of value from community-based resources, such as, among others, people, land, skills, culture, and reinvesting that value in a way that is sensitive to the socioeconomic development of communities (see, e.g., Colbourne et al., 2023; Colbourne, 2017; Peredo and Anderson, 2006). This has been demonstrated in several studies, such as among others that of the Bedouins who continued to engage in sheep farming without profit for the sake of maintaining Bedouin traditions and culture (Degen, 2007), and the Maasai in Kenya who were generally reluctant to be an integral part of the market economy with its emphasis on profit (Ndemo, 2007). In the case of Canada, the Namgis First Nation engaged in entrepreneurial ventures which were of symbolic significance, although of course there were some enterprises which aimed for economic gain (Dana, 2007). What also distinguishes indigenous entrepreneurship from mainstream economic activities is that they rely on locally available resources, which are generally used sustainably, unlike modern economic systems, which degrade and destroy the environment (Dana, 2015). Stated differently, Indigenous communities use IKS to innovate using resources in their environment, and given that they frequently operate in resource-constrained contexts, their inventions are typically simple and fundamental, yet efficacious (Onwuegbuzie and Mafimisebi, 2021). Through the use of IKS, Indigenous communities provide solutions to socioeconomic development and other challenges. These solutions are usable, context-specific, and grounded in preserving and enhancing the natural environment and its patterns and processes (see, e.g., Akbar et al., 2020). This points to the fact that IKS and the indigenous solutions it provides have the potential to foster sustainable economic development (see, e.g., Dana, 2015).

Although it is accepted that approaches to entrepreneurship may vary from one Indigenous community to another, such approaches are usually at variance with the fundamental tenets of Eurocentric economic theories, which emphasize profit. This is not surprising because indigenous entrepreneurships are based on a set of different cultural values, and their creation is not solely based on responding to market needs but also complex kinship, family, clan/community, and social ties (Dana, 2015). This is one of the major themes that are discussed in this book by drawing from case studies across the African continent. Therefore, although "mainstream economics suggest that rationally one might choose to maximize profit, we learn from Indigenous people that entrepreneurship also has non-economic causal variables. The individual profit motive exists; however, there are also community needs and objectives," which form the definition and measures of success (Dana, 2015:160). Another way to look at this is that financial viability is not the focus or target in and of itself; rather, it is a prerequisite for achieving other outcomes, such as among others, preservation and respect for traditional values and customs, poverty alleviation, better living conditions, and employment creation (Dana, 2015). Simply put, "value to society is not synonymous with value to a firm and monetary value is not necessarily social value" (Dana, 2015:165). All this taken together suggests that in indigenous entrepreneurship, the definition and measurement of success is different from that in the mainstream market economy because the focus is on community development, cultural preservation, environmental conservation, and sustainability inter alia.

In this regard, this book advances the argument that indigenous entrepreneurship provides a viable and more sustainable pathway to inclusive and wholistic socioeconomic development in Africa. The inclusivity and wholism derive from the fact that while indigenous entrepreneurship does focus on profit, this is not at the expense of but also focuses, if not more, on community development, cultural preservation, and environmental protection. Consequently, without romanticizing Indigenous entrepreneurship in Africa, but seeing it this way illuminates two issues. First, it frees it from the Eurocentric definition, which assumes that profit is the objective in all types of entrepreneurships in Africa. Freeing the definition and understanding of indigenous entrepreneurship from the universalizing tendencies of Eurocentrism and Euromodernity constitutes a decolonial position on which this book is anchored. Second, it opens up the possibility of alternative ways of thinking about and pursuing socioeconomic development in Africa in transcendental ways. If this is the case, it is therefore possible to examine how this can be achieved in Africa, and this is one of the themes that animates this book.

From the definition of IKS and indigenous entrepreneurship, it should be evident that the latter is based on or is possible because of the former. In other words, IKS is a body of knowledge in its own right that has been utilized by people for Indigenous entrepreneurship, livelihoods, and socioeconomic development, among others. In the context of the ongoing decolonization debate in Africa, it is imperative to bring into the conversation the issue of IKS and entrepreneurship in Africa in terms of how the conceptualization and praxis of indigenous entrepreneurship can be unlocked and unleashed to achieve socioeconomic development in Africa. This is pertinent because, as suggested above, Indigenous entrepreneurship is different from mainstream economic activities, even though it has some elements of them. Given the limitations in the conventional idea of entrepreneurship, there is a need to reimagine and reinterpret entrepreneurship to fit into the existing knowledge systems in Africa. It should be remembered that before colonial conquest, indigenous entrepreneurship flourished and led to authentic and inclusive socioeconomic development (Ochonu, 2018). It was colonialism and settler Eurocentric institutions that subjected indigenous entrepreneurship to mainstream economic governance regimes, leading to its marginalization and, in some cases, relative demise (see, e.g., Colbourne et al., 2023). Nonetheless, given on the one hand the resilience of indigenous entrepreneurship in terms of surviving the colonial encounter, onslaught, and institutions that should have suffocated and buried it, and on the other its potential to spur inclusive socioeconomic development, it is now time to revisit and recover it without subjecting it to or seeing it through mainstream economic lenses and the Eurocentric frame. More than that, entrepreneurship has frequently been offered as a potential option for reconstructing Indigenous communities, many of which have been bruised by colonization and the accompanying degradation of traditional philosophical, social, ecological, linguistic, and cultural resources and identities (Murphy et al., 2020).

Consequently, this raises three critical but related questions, which the book tackles: first, how is entrepreneurship understood and practiced by Indigenous communities in Africa? Second, why is there a need to decolonize the conceptualization and praxis of entrepreneurship in Africa through the logic of IKS, and how can this be done? Third, how and to what extent can IKS be leveraged and/or mobilized and unleashed to contribute to socioeconomic development in Africa? Against this backdrop, the main goal of the book is to decolonize entrepreneurship within the prism of IKS and its application to sustainable livelihood. This task should show the poverty of the current understanding and implementation of entrepreneurship in Africa, which is couched within a Eurocentric frame. It is now time to transcend the Eurocentric monologue of entrepreneurship to an understanding of how Indigenous communities in various parts of Africa conceptualize and practice entrepreneurship, leading to socioeconomic development. It is important to amplify the point that, since it is the cornerstone of the contemporary indigenous economy, indigenous entrepreneurship is both the answer and the path forward for the development and cultural survival of African communities. It is important to remember that Indigenous communities in Africa, Latin America, North America, and other regions of the world have a long history of business ownership, innovation, and operations, including, among others, engaging in barter trading (Cheteni and Umejesi, 2024; Munawar et al., 2020).

Decoloniality

Decoloniality is the theoretical axis that anchor this book because of "its push for shifting of geography of reason from the West as the epistemic locale from which the 'world is described, conceptualized and ranked' to the ex-colonized epistemic sites as legitimate points of departure in describing the construction of the modern world order" based on three analytical concepts which are the coloniality of power, coloniality of being, and coloniality of knowledge (Ndlovu-Gatsheni, 2013:13). For these reasons, this book pushes for the decolonization of entrepreneurship in Africa such that it is seen and framed within the precolonial knowledge system that can lead to and support viable enterprises for inclusive socioeconomic development. In this sense, decoloniality "resists or is against colonial domination. It is not just resistance or a posture of resistance, but it's for the ongoing creation of ways of thinking, of ways of knowing, of ways of sensing, being, and living outside coloniality. Outside or despite coloniality, and in its borders, its fissures, and its cracks. So, in that sense, decoloniality necessarily brings forth or points to the issue of coloniality" (Walsh, 2021: 1). Coloniality of power describes the unequal power relations between the global South and Western European and North American countries, which are divided by an imaginary abyssal line. This abyssal line places Western European and North American countries in the Zone of Being, whereas those in the global South are classified as belonging to the Zone of Non-Being (see Santos, 2007). This has provided the basis for abysmal thinking (see Santos, 2007) and the justification of, among others, slavery and exploitation by Western European and North American countries of those countries in the global South (see, e.g., Ndlovu-Gatsheni, 2015). The interest of this book on the notion of the coloniality of power rests on its utility in showing that power rests with Western European and North American countries, and this is why Euromodernity has been universalized (see, e.g., Ndlovu-Gatsheni, 2015, 2018, 2020), including leading to misrepresentations of what Indigenous entrepreneurship is and what it can or cannot do. In this book, we contest this by bringing to the fore the value and importance of indigenous entrepreneurship as an important pillar in the socioeconomic development of Africa. Coloniality of being describes the colonization of the human itself (Maldonado-Torres, 2007) or what Wynter (2003) refers to as the invention of the human. Such an invention has constructed Whiteness as a symbol of the human and the rest as subhuman.

In this book, we deploy the notion of coloniality of being to argue that there is nothing subhuman about Indigenous communities and their IKS in Africa. For this reason, such Indigenous communities should be seen as possessors of viable and important knowledge, which, if properly understood and harnessed, can provide an important pathway toward inclusive socioeconomic development in Africa. For example, what entrepreneurship means to Indigenous communities in Africa should not be dismissed but understood as another way of conceptualizing and implementing socioeconomic development. This amplifies the notion of the coloniality of knowledge which concerns "the politics of knowledge generation as well as questions of who generates which knowledge, and for what purpose" (Ndlovu-Gatsheni, 2013:11). In this regard, it should be remembered that "Western enlightenment thought has [...] posited itself as the wellspring of universal learning, of Science and Philosophy [...], it has regarded the non-West—variously known as the ancient, the orient, the primitive world, the third world, the underdeveloped world, the developing world, and now the global south—primarily as a place of parochial wisdom, of antiquarian traditions, of exotic ways and means" (Comaroff and Comaroff, 2012:1). The coloniality of knowledge assists this book to contest the Eurocentric definition of entrepreneurship as the only knowledge and which marginalizes or confines Indigenous entrepreneurship to the category of nonexistence. It assists in showing that the marginalization of Indigenous entrepreneurship has been one of the strategies to promote mainstream economic activities without any competition from the latter (see Dana, 2015). In this book, we argue that Indigenous entrepreneurship can also be one of the ways through which socioeconomic development in Africa can be achieved. There is no attempt to suggest replacing mainstream economic activities with Indigenous entrepreneurship, but to propose that there is another viable knowledge for entrepreneurship and socioeconomic development in Africa beyond the Eurocentric frame, which has been silenced, and now is the time to unleash it. In this sense, the theoretical purchase of decoloniality in this book is that it:

> is a concept, it's an analytic, but it's also, which is particularly key for me, a form of praxis, which is increasingly assumed by social movements, by communities, by engaged intellectuals, by artists, by activists and many, many others against the ongoing violence, dispossession, and war altogether waged against specific bodies, against people's cultures, knowledges, spiritualities, and against nature, and for the insurgent and resurgent creation, construction, and possibility of other modes of that knowledge being existence in life (Walsh: 2021:1).

By deploying decoloniality, we argue that the complacent position of Eurocentrism in entrepreneurship must be "confronted, and a different thought constructed and positioned from the histories and subjectivities of the people" (Walsh, 2012:11).

Structure of the Book

This book is composed of eight chapters. This introduction is followed by Chapter 2, which expands the theoretical grounding and/or framing of the book by engaging with the debates around the decolonization of knowledge and the epistemology of entrepreneurship in African Indigenous communities. It argues that many Indigenous communities in Africa do not subscribe to the Eurocentric idea and philosophy of entrepreneurship, which prioritize profit maximization and wealth creation. Quite to the contrary, indigenous entrepreneurship advances community development. For this reason, the chapter posits that there is a need to decolonize the idea and practice of entrepreneurship so that it is inclusive of Indigenous communities. It is only after such decolonization has occurred that entrepreneurship that is informed by and responds to the Indigenous communities can make socioeconomic development possible and meaningful. This point is illustrated in Chapter 3, which unmasks the poverty of the formality–informality dichotomy in the context of Indigenous entrepreneurship. Chapter 3 thus takes forward the discussion on decolonizing entrepreneurship by interrogating the problem of classifying or constructing Indigenous entrepreneurship as informal and therefore unimportant. The chapter argues that there is therefore a coloniality in the formality–informality dichotomy, which is a technique of coloniality that continues to stifle, suppress, and depoliticize not only informality in general but IE.

This chapter posits that framing economic activities that contribute significantly to the Gross Domestic Product and create jobs for more than 50 percent of the population in some instances, are rooted in colonial epistemologies that misrecognize the structural and productive role of African indigenous economic systems. Drawing on political economy, postcolonial and decolonial theory, and comparative case studies of Nigeria, Kenya, South Africa, and Rwanda, it advances a critique of the language of informality and proposes pathways for transforming what is generally regarded as informal activities into viable, inclusive, and sustainable economic systems. The chapter demonstrates that meaningful transformation requires not only regulatory reform but also epistemic shifts, institutional innovation, and co-produced policy frameworks grounded in African realities. It argues that rethinking the language of economic policy must be preceded by a deliberate de-centering of Western economic models and a recognition of Indigenous entrepreneurship as a legitimate and generative form of economic practice. Such an approach enables the contextualization of development, expands the analytical lens through which African economies are understood, and repositions Indigenous entrepreneurial practices as central rather than peripheral to economic policy and transformation. Simply put, it is time for Indigenous entrepreneurship not to be seen as informal and an aberration, but an integral component of the economic activities, if the goal of inclusive, sustainable socioeconomic development in Africa is to be achieved.

To deepen the discussion of how IKS are important in matters of self and community care as well as resource use and governance, chapter four discusses the extent to which these practices can be leveraged within a decolonial praxis, for community and socioeconomic development in Africa. This chapter examines how entrepreneurship opportunities can be better designed and managed through the reform of governance mechanisms among the Nguni Indigenous people of South Africa. This chapter contributes to the discourse on Indigenous entrepreneurship by highlighting contemporary governance gaps in youth entrepreneurship policy responses and advocating for inclusive employment opportunities at the community level.

During and postcolonial conquest, Indigenous entrepreneurship has proven to be resilient in the face of adversity and marginalization. For example, many Indigenous entrepreneurs in Africa face financial obstacles in terms of accessing finance, and yet this has not stopped the small business operations of Indigenous communities. In light of this, Chapter 5 invokes the case study of *Esusu* (a customary method of collaboration among the Yoruba people of Nigeria, which involves groups of people who make contributions to credit and savings organizations for their mutual gain) to illustrate how and to what extent Indigenous entrepreneurs finance their enterprises. *Esusu*, which is a form of daily/weekly/monthly contributions by women in the market, is usually done to raise money either to establish or expand a business, and this has assisted Indigenous entrepreneurs in mitigating the challenges of access to finance on the African continent. Seen thus, *Esusu* is an alternative financial system that Indigenous entrepreneurs have devised to circumvent the elitist structure of the formal financial system. This needs to be and can be expanded and supported such that Indigenous entrepreneurship in Africa can grow and expand, leading to socioeconomic development.

Beyond the recognition and expansion of Indigenous methods of financing entrepreneurs is the related question of the gender dimension in Indigenous entrepreneurship in Africa. The point is that there is often a silencing and underrepresentation of the gender question in Indigenous knowledge. This is concerning, considering that women in Africa contribute substantially to the development of Africa. In addition, the discourse around Indigenous entrepreneurship is at best constructed as informal and at worst as illicit and/ or illegal, with the result that there is a lack of their recognition. This is often worse for women entrepreneurs based on their double status as women and entrepreneurs. This is a decolonial issue based on the need to, first, contest and ultimately dismiss the very idea that Indigenous entrepreneurship is illicit. Second, there is the need to recenter the contribution of women to Indigenous entrepreneurship and socioeconomic development in Africa. This is the context within which chapter six brings in a gendered dimension to the discussion on Indigenous knowledge in Africa, by illustrating that women have been major stakeholders in Indigenous entrepreneurship in Africa. Informed by a decolonial position, the chapter calls for the mainstreaming of women in Indigenous entrepreneurship in Africa and an obliteration of androcentrism in the same.

One of the major arguments of this book is the need for the recognition of IKS and Indigenous entrepreneurship, leading to ecologies of knowledges and ontological pluralism (Santos, 2007). This is "informed by a strong conviction that all human beings are not only born into a knowledge system but are legitimate knowers and producers of legitimate knowledge" (Ndlovu-Gatsheni, 2018:24). If this is the case, which it is, Chapter 7 argues that there are values and resources in African Indigenous knowledge that can be harnessed to foster climate adaptive entrepreneurship in Africa. From bio-energy transition to bio-agriculture, Indigenous knowledge can help to mitigate the challenges of climate change in Africa through more conscious protection of the environment and drawing on the experiences and lessons of the past. This is important and urgent because climate change has become an incubus on the global economy. In Africa, the problem of climate change has become intractable and constitutes a strong barrier to the ongoing desire to industrialize. From energy generation to agriculture, the conventional approaches to addressing climate change have proved to be counterproductive, and this points to the relevance of IKS in the epistemic domain, such that there are ecologies of knowledge (Santos, 2007) to successfully spur Africa's socioeconomic development challenges.

To deepen this argument, Chapter 8 illustrates a specific case in which Indigenous entrepreneurship has led to socioeconomic development in Africa. This chapter examines the role of Indigenous entrepreneurship (IE) in advancing sustainable livelihoods in Africa. Drawing on empirical evidence from selected Indigenous communities in South Africa, it shows that Indigenous entrepreneurial practices are embedded in community-oriented values that prioritize care, reciprocity, and collective well-being and ecological stewardship rather than narrow profit maximization.

The chapter argues that such models challenge dominant market-driven approaches by fostering inclusive development, strengthening local economies, and reinforcing social cohesion. Using a sustainable livelihoods perspective, the analysis demonstrates that IE is not an isolated economic activity, but a relational process grounded in IKS and social networks. These practices support collective decision-making, enhance resilience, and sustain local resource systems. The chapter illuminates IE's capacity to generate development outcomes that are context-specific, socially embedded, and environmentally sustainable. This positions IE as a viable pathway for locally grounded development, particularly in contexts where conventional, externally driven models have proven inadequate or exclusionary. While the findings confirm that IE strengthens multiple forms of capital, they also reveal that these processes unfold within conditions of structural inequality, environmental stress, and institutional exclusion. The chapter further argues that IE both operates within and stretches the SLF by demonstrating that livelihood assets in Indigenous contexts are not discrete but co-produced through social relations, cultural norms, and environmental practices. In doing so, it challenges asset-based interpretations of livelihoods and calls for a more relational and politically grounded reading of the SLF that accounts for power, history, and Indigenous epistemologies.

Another dimension to the marginalization of IKS in Africa is in the policy domain. This is because the policy on entrepreneurship in Africa has been mostly based on Eurocentric knowledge. In many instances, these policies were designed without the voices of the intended beneficiaries and are based on templates designed in Western and North American countries. Therefore, Chapter 9 argues that given the limitations and failures of the Eurocentric approaches to designing policies on Indigenous entrepreneurship, a decolonized approach that is grounded in ownership and co-design of the policies by Indigenous communities in Africa is pertinent. This could be one of the ways that can unlock and unleash the socioeconomic development potential of Indigenous entrepreneurship in Africa.

This argument is taken further in the concluding chapter of this book, in which we advocate for the leveraging of IKS toward a new socioeconomic development agenda in Africa. In this chapter, we problematized development beyond economic growth, insisting that development transcends growth in Gross Domestic Product to include human dignity, cultural pride, respect for language and identity. We also showed the various areas in which indigenous knowledge can be applied to foster sustainable livelihood in ways that incorporate concerns for the environment, build trust among community members, mentor a new generation of entrepreneurs, and enlarge the scope of opportunities for economic activities in Africa. We amplify the argument that there seems to be a critical disjuncture in how entrepreneurship is understood and practiced among Indigenous communities in Africa and how it is framed and practiced within the Eurocentric logic.

For this cause, it is advanced that it is now time to transcend the Eurocentric monologue of entrepreneurship to an understanding of how Indigenous communities in various parts of Africa conceptualize and practice entrepreneurship, leading to socioeconomic development. This is premised on the relegation of the value and importance of IKS as a body of knowledge that can be unleashed and, in the process, provide solutions to many socioeconomic problems affecting communities in the Global South, particularly in Africa, has been underscored. In other words, there has been a recovery of the importance of IKS in socioeconomic development buried in the colonial cemetery. This is the context within which this book volume deploys the notion and praxis of IKS to decolonize entrepreneurship and propose a potential re-conceptualization of socioeconomic development in Africa. The book provides a comprehensive understanding of the intersection between Indigenous knowledge and development in Africa; it touches on different thematic areas relevant to the discourse. The book's strength lies in its complementary efforts to previous works at establishing how indigenous entrepreneurship can positively impact Africa.

References

Akbar, N., Abubakar, I. R., & Bouregh, A. S. (2020). Fostering urban sustainability through the ecological wisdom of traditional settlements. *Sustainability, 12*(23), 1–19.

Cheteni, P., & Umejesi, I. (2024). Indigenous knowledge in entrepreneurship and cultural tourism in the rural areas. *Comparative Sociology, 23*(3), 340–359.

Colbourne, R. (2017). Indigenous entrepreneurship and hybrid ventures. In A. C. Corbett & J. A. Katz (Eds.), *Hybrid ventures* (pp. 93–149). Emerald Publishing.

Colbourne, R., Peredo, A. M., & Henriques, I. (2023). Indigenous entrepreneurship? Setting the record straight. *Business History, 66*(2), 455–477.

Comaroff, J., & Comaroff, J. L. (2012). *Theory from the South: Or, how Euro-America is evolving toward Africa*. Paradigm.

Dana, L.-P. (2007). The Namgis First Nation of Alert Bay, Cormorant Island, British Columbia. In L.-P. Dana & R. B. Anderson (Eds.), *International handbook of research on indigenous entrepreneurship* (pp. 328–335). Edward Elgar.

Dana, L.-P. (2015). Indigenous entrepreneurship: An emerging field of research. *International Journal of Business and Globalisation, 14*(2), 158–169.

Degen, A. A. (2007). Traditional livestock production among Bedouin in the Negev Desert. In L.-P. Dana & R. B. Anderson (Eds.), *International handbook of research on indigenous entrepreneurship* (pp. 115–136). Edward Elgar.

Maldonado-Torres, N. (2007). On coloniality of being: Contributions to the development of a concept. *Cultural Studies, 21*(2–3), 240–270.

Mawere, M. (2015). Indigenous knowledge and public education in Sub-Saharan Africa. *Africa Spectrum, 50*(2), 57–71.

Mwangi, E. (1998). *Colonialism, self-governance and forestry in Kenya: Policy, practice and outcomes.* Indiana University.

Moyo, I. (2022). Beyond a tokenistic inclusion of indigenous knowledge systems in protected area governance and management in Okhahlamba-Drakensberg. *African Geographical Review, 42*(2), 141–156.

Munawar, A., Duwila, U., & Harini, D. (2020). A review on types of money in the development of community trade. *Journal of Critical Reviews, 7*(8), 1993–1997.

Murphy, M., Danis, W. M., & Mack, J. (2020). From principles to action: Community-based entrepreneurship in the Toquaht Nation. *Journal of Business Venturing, 35*(6).

Ndemo, B. (2007). The Maasai: Entrepreneurship and change. In L.-P. Dana & R. B. Anderson (Eds.), *International handbook of research on indigenous entrepreneurship* (pp. 84–99). Edward Elgar.

Ndlovu-Gatsheni, S. J. (2013). Why decoloniality in the 21st century? *The Thinker, 48*, 10–15.

Ndlovu-Gatsheni, S. J. (2015). Decoloniality in Africa: A continuing search for a new world order. *ARAS, 36*(2), 22–50.

Ndlovu-Gatsheni, S. J. (2018). *Epistemic freedom in Africa: Deprovincialization and decolonization.* Routledge.

Ndlovu-Gatsheni, S. J. (2020). *Decolonization, development and knowledge in Africa: Turning over a new leaf.* Routledge.

Ochonu, M. (Ed.). (2018). *Entrepreneurship in Africa: A historical approach.* Indiana University Press.

Onwuegbuzie, H. N., & Mafimisebi, O. P. (2021). Global relevance of scaling African indigenous entrepreneurship. *Technological Forecasting and Social Change, 166*, 1–16.

Peredo, A. M., & Anderson, R. B. (2006). Indigenous entrepreneurship research: Themes and variations. In C. S. Galbraith (Ed.), *Developmental entrepreneurship: Adversity, risk, and isolation* (pp. 253–273). Emerald Publishing.

Santos, B. de S. (2007). Beyond abyssal thinking: From global lines to ecologies of knowledge. *Review, 30*(1), 45–89.

Senanayake, S. G. J. N. (2006). Indigenous knowledge as a key to sustainable development. *Journal of Agricultural Sciences, 2*(1), 87–94.

Walsh, C. (2012). "Other" knowledges, "other" critiques: Reflections on the politics and practices of philosophy and decoloniality in the "other" America. *Transmodernity, 2*(1), 11–26.

Walsh, C. (2021, May 3). What do we mean by "decoloniality"? A discussion (Walter Mignolo, Catherine Walsh, Tink Tinker, Fernando Herrero). *The New Polis.* https://thenewpolis.com/2021/05/03/what-do-we-mean-by-decoloniality-a-conversation/

Chapter 2

DECOLONIZING KNOWLEDGE ON YOUTH EMPLOYMENT AND ENTREPRENEURSHIP IN AFRICA: WHY THE VOICES OF INDIGENOUS YOUTH MATTER?

Introduction

The challenge of youth unemployment is a staggering one that continues to stimulate debate among scholars, policy makers, and development workers (Mwakalila, 2022, World Bank 2024). The youth bulge in Africa represents both a challenge and an opportunity. According to the African Development Bank (2016), with over 60% of its population under 25 years of age, Africa faces a very strong pressure to provide employment and entrepreneurial opportunities. As Oloruntoba, Moyo, and Zondo (2025) argue, the orthodox approach based on Western models to addressing these inscrutable challenges has not succeeded. The lack of success can be linked to the neglect of the Indigenous Knowledge Systems (IKS) upon which most of the communities were established and on which they have lived for generations. This chapter posits that decolonizing knowledge, which includes reclaiming indigenous epistemologies and practices, is essential for building inclusive, sustainable pathways for youth employment and entrepreneurship in Africa.

One of the consequences of colonialism was the deliberate silencing of African knowledges (Mawere, 2015). Hiding under the mission of civilizing the so-called natives and ensconced within the euphoria of universalism of Eurocentric knowledge, the colonialists committed epistemicides (Ndlovu-Gatsheni, 2018, Mignolo 2009). Through political, economic, social, and psychological reorientation and historical distortions, the colonialists ensured the subordination of African endogenous knowledges in various aspects of life such as food preservation, health, environmental conservation, conflict resolution, and so on (Hountondji, 1996). Despite some attempts made by the postcolonial state to decolonize Eurocentric knowledge through the establishment of universities and other centers of learning (Oloruntoba, 2014), coloniality subsists in the ways in which African traditional and indigenous knowledge are still being considered as inferior to Western knowledge and science. Marginalized communities in Africa suffer from the double jeopardy of exclusion during colonial and postcolonial times. Due to their attachment to nature and lack of interest in engaging with the postcolonial society, marginalized groups like the Maasai of Kenya and Tanzania, Ogiek of Kenya, Basongora and Batwa of Uganda, Kenya, Malawi, Rwanda, Tanzania, and Uganda, Batwa, Khoi and San of South Africa and Botswana, Ijaw and Ogoni of Nigeria suffer from further exploitations and exclusion by dominant groups (Nyiayaana, 2025, Okolo, 2025). In instances where the postcolonial state has intervened to support economic livelihood among these groups of people, such interventions have been done without a systematic process of engagement with the affected communities in ways that can enhance the centering of their voices (Oloruntoba, 2025). Rather, the postcolonial state has adopted Western ideas of entrepreneurship, which in many instances are different from the idea of entrepreneurship shared by the marginalized communities. Whereas the western idea of entrepreneurship is based on individual accumulation, profiteering, and extraction of nature, marginalized communities have a community approach to entrepreneurship, which entails sharing of wealth, setting an agenda for business practices, and even mobilizing resources. Besides, Indigenous people also care for nature and seek ways to preserve it from degradation (Colborne, Peredo, and Henriques, 2023). According to Di Giorgo and Habibis (2019:32), there are both ontological and epistemological gaps that exist between the dominant policy approaches and indigenous understandings of work and employment. Although the case study of their analysis was Australia, this also applies to how the postcolonial state in Africa relates to the marginalized communities.

In this chapter, we argue that the silencing of the voices of the youth and community elders in the marginalized communities and the lack of application of the IKS of the people have affected the success of such policies in Africa. Thus, it is imperative to decolonize the knowledge on employment and entrepreneurship through a deliberate process of engagement with the indigenous communities and their intergenerational knowledge. Although some Eurocentric scholars have argued that entrepreneurship broadly conceived is not in tandem with African culture (Cohen, 1982), we challenge this narrative by arguing that in precolonial Africa, Africans engaged in different forms of micro to medium-scale enterprises like wood carving, clothing, arts, traditional medicines, rain making, fashion, and other forms of entrepreneurial activities. They also engage in long-distance trade through which they built networks of exchange (Ochonu, 2015). As Zeleza (1993) argues, in precolonial times, Africans engaged in different forms of economic activities that led to the building of empires and nation-states. This great achievement was violently disrupted through the slave trade, imperialism, colonialism, and increasingly coloniality (Rodney, 1981; Ake, 1981). For more discussions on coloniality and how it affects peripheral regions of the world, see Mignolo (2009). The various economic activities engaged in by marginalized communities, such as cattle rearing, fishing, farming, clothing, and a host of others, are valid entrepreneurial activities that have sustained these groups of people for generations. The colonial enterprise delegitimized the economic systems of the Indigenous Peoples by introducing new economic systems that support their exploitation and intense search for revenue to service their home economies.

Scholars have argued that much of what constitutes Western knowledge, under which African indigenous knowledges have been subsumed, contains vast knowledge either stolen or violently taken from Africa by the colonialists (see Mawere, 2015). As Mbembe (2021) argues, most of the artifacts stolen from Africa during the colonial invasions of the late nineteenth century and early twentieth century were not mere objects. They contain vast knowledge, spirituality and part of the beings and identity of Africans who designed and preserved those materials for centuries long before the violent encounters with the colonialists. The contradictions of dispossession and denial of entrepreneurship knowledge among Africans is not limited to the people of the continent. Peredo and Henriques (2023) argue that in the case of Canada, Indigenous People were not only engaged in entrepreneurship activities before the arrival of the settler colonialists but initially collaborated and supported the latter when they arrived. These scholars debunked the claim that Indigenous People did not engage in entrepreneurship activities before the colonial intrusion into the country.

Unlike Canada where efforts are being made by the government to reconcile with Indigenous People and indeed support various entrepreneurship initiatives, the silencing of African indigenous knowledges and lack of engagement with marginalized communities continue apace due to lack of decolonization both at the national and subnational levels. Lack of application of indigenous knowledges in the design of employment and entrepreneurship in postcolonial time has affected the extent to which these policies have succeeded. This chapter further examines the current debates on decolonization of knowledge and how the application of African indigenous knowledge systems can bring about a decolonial turn in the various interventions toward addressing the problems of youth unemployment and limited success of entrepreneurial endeavors among the indigenous groups in Africa.

The rest of this chapter proceeds as follows. The next section presents the analysis of the politics of indigenous knowledge and the contentions over its relevance, especially in the light of the historical experiences of colonization and coloniality in the post-colonial state in Africa. The third section examines various interventions of the postcolonial state to promote economic activities among indigenous groups in Kenya, South Africa, Tanzania and Uganda and how these attempts have failed because of the lack of or limited engagement with indigenous communities. In this section, we will show that rather than engaging with Indigenous groups to know how their knowledge can be harnessed to promote entrepreneurship, they have rather invited multinational corporations which ended up displacing and dispossessing some of the Indigenous communities of their ancestral lands. In section four a strong case for decolonization of knowledge, especially on entrepreneurship through centering of the voices of indigenous groups in Africa is made, while section five concludes.

The Politics of Knowledge Production and the Contentions over Africa's Indigenous Knowledge

Knowledge production is a natural process of human existence from time immemorial. Societies and communities exist based on knowledge that they either inherited, adapted or generated to meet their immediate political, economic, environmental, health and social needs. Knowledge generated by societies have contributed to their civilizations and guided their existence over time. As mentioned above, Africa was not excluded from this experience as knowledge generated in the continent led to building of powerful kingdoms, empires, and creation of arts, many of which are in museums in western world today (see Mbembe 2021). Various scholarly accounts have showed how knowledge from Africa laid the foundation of what is known as Greece civilization thousands of years ago (Mawere, 2015). However, due to series of violent encounters with Europe and Arabs, such as slave trade, imperialism and colonialism, a systematic and deliberate attempt was made to deny the authenticity of African knowledges, (Mawere, 2015). Basking from the scientific and military breakthroughs that followed the Enlightenment in Europe, the intellectual and political elites assumed a superior position over other races and provincialized knowledge under what they called universalism (Ndlovu-Gatsheni, 2018). Rather than being humble to recognize the pluriversal nature of knowledge in which different societies have their own knowledge systems, Eurocentrism was reified and forced on the conquered territories in the Americas, Asia and Africa.

They did this through vociferous denial of the knowledges of others, reduction of their languages to vernacular and reconstitution of their identities to barbarians who need civilization. If some early European anthropologists and philosophers laid the foundation for the narrative of occupying a heart of darkness (Conrad, 1896), colonialism provided the official structure through which the culture, knowledge, language and the history of the natives were subordinated (Memmi, 1991). The use of formal institutions to silence, otherize and negate the authenticity of the knowledges of Indigenous communities led to epistemicides (Mignolo, 2009). Epistemicides cuts into the very being of the colonized as it robs them of their language, culture, self-worth and identity. Of particular importance is language which the colonialists replaced and introduced their own as the main language of instruction. As Wa thiong'o (2005) argues, using the language of the colonizer as a medium of instruction affects learning and undermines the identity of the colonized, hence his insistence that writing in one's original language is the best way of expression. In relations to youth entrepreneurship, the use of colonial language has affected the extent to which economic policies and programs are framed and communicated from the colonial to the postcolonial terms. For instance, scholars have argued that the use of the word informal to describe economic activities such as roadside trading, cross border trade, artisans etc. is a product of coloniality and structural subordination of Indigenous economic systems in Africa with those of the colonialized (Oloruntoba and Akinboye, 2025; Olaitan, 2024).

The form of education provided by the colonialists led to miseducation in which Africans who attended those schools were made not to think critically but to see themselves as inferior to the colonialists. Through its policy of assimilation, France went as far as taking pride in making Frenchmen out of the natives in what is now regarded as Francophone Africa (Adebajo and Whiteman, 2012). The usurpation of the existing structures of education, knowledge and being was a clear demonstration of the misuse of power against the colonized. To enforce the superiority of the colonial language, local languages were considered inferior and disallowed as a means of communication. Economically, the precolonial economic system based on household subsistence agriculture and anchored on African indigenous knowledge systems was supplanted and in its place a wage economic system was introduced (Ake, 1981). This form of economic system continues into the post-independent era. Although there is no single approach to how the state has designed economic policies, in many instances, such policies have been designed in line with the logic of western-oriented capitalist economic system. Based on the classical economic argument that the newly independent countries lack the required capital to build their economies, official aid and assistance were not only provided but multinational companies were supported to establish branches of their companies in several African countries (Easterly 2006). These companies, which have traditionally been in mining and extractive sectors have done little to improve the living conditions of Africans. In mining communities, environmental degradation continues apace, with unsupported and unregulated artisanal mining. Climate change has become worsened due to the neglect of the indigenous knowledge systems used by the people to preserve the environment in precolonial times.

Although experiences are varied in the ways in which the postcolonial state relate with the marginalized communities covered in this study, they have more or less replicated the colonial forms of silencing of the voices of the marginalized people. In South Africa for instance, the unique identity of the indigenous Khoi and San people have not been recognized. Contrary to the United Nations Declaration on the rights of Indigenous People (United Nations, 2007), which recognized the rights of Indigenous Peoples to exist on their own terms and in accordance with their cultures, norms, and mores, the South African government would rather locate the rights of the people within the constitution. Despite the constitutional recognition of the rights of this Indigenous group, they remain one of the groups with the least access to finance to carry out their entrepreneurial activities. Worse still, the land on which they have their herbs and other forms of livelihood are either leased out to big corporations or kept as reserves, by so doing effectively denying the people access to these resources (Indigenous World, South Africa, 2024).

The contradictions of constitutional recognition and continuous systematic land dispossession raise an important question around the extent to which the voices of the marginalized people were incorporated in the process of drafting the constitution. There are also questions on the extent to which the constitution protects the Indigenous People, particularly in relations to their rights to use their resources to improve their economic conditions and livelihoods. Land dispossession goes with knowledge dispossession because for Indigenous People, the land and the nature are generators of knowledge. Lack of recognition and application of indigenous knowledge robs South Africa of opportunities to use the rich indigenous knowledge in the marginalized communities to establish or scale up their enterprises and create jobs in the process.

Within South Africa, poverty and employment are higher in the provinces where the Indigenous People are located than in the rest of the country. The inability of the postcolonial state in Africa to duly recognize and accord Indigenous Peoples the respect that they deserve has not allowed for a full engagement with their knowledges in ways that can lead to a win–win outcome both in the preservation of those knowledges and in harnessing the opportunities inherent in them for livelihood and sustainable economic development. In Tanzania, the government has not recognized Indigenous People as people with distinct identity. This lack of recognition had its root in the colonial display of power in which they divided the Maasai people between Kenya and Tanzania. The Maasai people in Tanzania just as in those in Kenya, have faced land confiscation and dispossession. Lack of access to political power and representation by the Maasai, alongside other Indigenous groups in Tanzania has resulted in a decline in their land, as well as in their cattle rearing and sustainable livelihood. Whereas the conservation of the environment matters very much to the Maasai people, there have been a strong argument in parliament over the decision of the government to clear Loliondo and Ngorongo and relocate the Maasai and Hadzabe communities (The East African, June 26, 2022). The relocation of these Indigenous People from their ancestral lands was allegedly done for the conservation of wildlife, create room for hunting for a Dubai based company, as well tourism (The East African, June 26, 2022).

While the government has the constitutional authority to decide on the best policy to increase economic growth, this should not be done at the expense of the right of the citizens to live in their ancestral land, to which they have been attached socially, spiritually, economically, and psychologically. The various forms of land dispossession and forceful relocation create conditions of poverty and unemployment among the Indigenous groups (International Working Group on Indigenous Affairs, 2019). What emerged from the cases above is the differences in how the postcolonial state uses its power in relation to the recognition of the unique identity of the marginalized groups. Giving due recognition to these groups of people through engagement and recentering of their voices can provide a great opportunity for the co-creation of knowledge on employment and entrepreneurship opportunities. In the next section, we analyze some of the various attempts by the African state to address socioeconomic challenges faced by the marginalized groups and how these have not yielded the desired result due to the lack of engagement with the voices of the indigenous youths, elders, and women in the marginalized communities.

In contrast to the approach taken by South Africa and Tanzania, Kenya has gone further to incorporate the unique identity of the indigenous people in that country in their constitution. The new constitution defines a "marginalized community" as one that "out of need or desire to preserve its unique culture and identity from assimilation, has remained outside the integrated social economic life of Kenya as a whole, or an indigenous community that has retained and maintained a traditional lifestyle and livelihood based on hunter or gatherer economy; or pastoral persons and communities whether they are nomadic or a settled community that because of its relative geographic isolation has experienced only marginal participation in the integrated social and economic life of Kenya as a whole." (Kenya Constitution, 2010). The constitution further requires the state to provide for adequate representation of "marginalized groups" in all levels of government, execute affirmative action on behalf of these groups, and promote the use of Indigenous languages and the free expression of traditional cultures (Cultural survival, 2010). The constitutional recognition of the rights of indigenous Peoples in Kenya is a positive change from the previous experiences of exclusion that the people have suffered. It is also a demonstration of the shift in the use of power in a postcolonial state. Although this is only a first step in the process of ensuring equal citizenship of Indigenous Peoples in Kenya, it signals great possibilities for reconciliation, restitution, and reintegration of the indigenous communities in ways that recognize their distinct identity.

Indigenous Entrepreneurship and State Interventions: Limits and Possibilities

Compared to other segments of the population, unemployment is higher among indigenous groups in different parts of the world. In Africa, the situation is even more dire because unemployment is generally high, especially among the youth. According to Ighobor (2017) figures from the World Bank show that youth accounts for 60 percent of all of Africa's jobs (World Bank, 2016). Young women are more affected with the problem of unemployment due to various sociocultural challenges that limit their ability to acquire skills. Youths from indigenous or marginalized communities are more affected with the plaque of unemployment given the nature of their existence and the challenges they continue to face with access to land, and the natural resources found in their environments. The governments of South Africa, Tanzania, and Kenya understand the importance of entrepreneurship in addressing the problem of unemployment. They have also developed various policies to address this problem.

In South Africa, for instance, government departments/ministries such as the Department of Small Business Development provide support for small- and medium-scale enterprises to thrive. The South African Micro Finance Apex Fund (SAMAF) also provides loans to entrepreneurs. There have also been discourses and policies on the importance of entrepreneurship, especially among the black population, many of whom are still trapped in poverty. This is because of the understanding that 'creating an entrepreneurial culture in South Africa is vital for economic growth and job creation. Also, entrepreneurship can drive innovation, increase productivity, and create new industries because it promotes a sense of self-reliance, creativity, and innovation, which are essential traits for any successful business (CABAN Investments, n.d). Despite the various programs put in place to promote entrepreneurship generally, the effect has been very limited (Tshikovh, 2021). Besides, there have been limited, if any, direct engagements between the government and the Khoi San people to promote indigenous entrepreneurship. The recent effort of the government to support the mass acquisition of land in the Western Cape to build a shopping mall over the land that the indigenous people consider to be sacred to them has been met with strong resistance. A court case has been instituted by representatives of the Khoi San people to halt the project (Global Citizen, 2021). While building malls can provide employment and increase revenue to the government through taxation, this goes contrary to the way of life of the marginalized people whose land is being forcefully taken over by the state. Building of malls is not only based on the indigenous knowledge of the people, but it is an approach based on imported knowledge on building a capitalist economy.

This concern over the preservation of historical heritage and conservation of the environment is at the root of the differences between how the postcolonial state and the marginalized people in South Africa perceive policies on addressing the problem of unemployment. Whereas it may be convenient for the government to argue that the marginalized people have access to the support provided to black entrepreneurs, the unique lifestyle and occupational practices of the Khoi San people constitute a serious barrier to their interest, if any, in accessing such support. What is needed is a humble approach by the state and the private sector to engage with the indigenous people on how to use their knowledge to create entrepreneurship programs that are in synchronization with their age-long practices, such as hunting, cattle rearing, and agriculture. In an age where climate change poses a serious challenge to the existence of all humanity, the indigenous knowledge of the Khoisan people on the preservation of the environment can contribute to addressing the existential challenges that we all face (Goats and Soda, 2014). In a period where big pharmaceutical companies are making drugs whose prices are outside the reach of the ordinary person, the herbal knowledge of the Khoisan people could contribute to making drugs that are easily affordable for the common people.

In Tanzania, the government has formulated various policies to address the twin problem of poverty and unemployment. For instance, in 2017, the government adopted a National Economic Empowerment Council (NEEC) that designed the Tanzania Inclusive National Entrepreneurship Policy (2017:1). The mission is stated as follows:

> To build the national entrepreneurship ecosystem, for enhanced entrepreneurial spirit and performance of economic activities, the creation of new start-ups and the formalization and growth of domestic MSMEs, upgraded productive capacities, as well as innovative and sustainable solutions to environmental and social challenges

The policy was very comprehensive as it incorporated the need to provide a legislative framework to guide entrepreneurial activities in the country, as well as the finance and training required to achieve the goals. However, this policy was based on the conventional knowledge of entrepreneurship. The rich tapestry of indigenous knowledge that underpins economic activities among the various marginalized societies was not reflected. The policy stated that in Tanzania, there is a high potential for entrepreneurship to become a channel for the social and economic integration of the most vulnerable groups. It also highlighted that to achieve the economic empowerment and effective participation in social progress of all citizens, there was a need to include youth, women, the rural poor, and other disadvantaged groups. The recognition of disadvantaged groups is very encouraging. However, almost six years after coming into effect, it remains to be seen if there have been any improvements in the living conditions of the marginalized communities in the country. The Maasai people possess very rich indigenous knowledge on various aspects of economic activities, which could have been harnessed to design a policy that will help in addressing the problem of unemployment among the group.

Beyond the top-down approach that has defined the ways in which employment and entrepreneurship policies have been designed, a co-creation of knowledge through the centering of the voices of the youth, community elders, women, and other stakeholders could yield better outcomes. An engagement with these categories of people falls within the purview of what constitutes indigenous entrepreneurship. As Colbourne (2017) argues, Indigenous entrepreneurship is a process of drawing value from community-based resources (people, land, capabilities, culture, etc.) and contributing value back that is responsive to a community's particular set of socioeconomic conditions. Thus, an engagement with the Maasai people in the design of entrepreneurship policies will not only ensure that their voices are heard, but such an approach will also ensure ownership.

Besides, the resultant strategy can lead to more positive outcomes because it will fit with the values, expectations, and priorities of the communities. This point is important because, as mentioned above, there is a fundamental difference in the epistemological foundation of entrepreneurship between dominant groups and marginalized communities. Whereas the dominant approach to and motive for entrepreneurship is individual profit-making and accumulation, marginalized groups have more benign goals that encapsulate a culture of sharing, community ownership, joint decision-making, and concerns for the environment.

A deliberate policy of engaging with the Maasai group, for instance, could create conditions for incorporating their knowledge on bead making, cattle rearing, hunting, and agriculture into such policies in ways that can ensure sustainability. The reconstitution of the dynamic of power in the relationship between the state and the marginalized groups is imperative to foster inclusion. As the challenges of unemployment continue to bite in Tanzania, opportunities exist to leverage existing indigenous knowledge among the Maasai to co-create policies and programs that can create jobs. It is important to emphasize that an intentional engagement with the Indigenous knowledges of the people will not only lead to employment among the people but in the country as a whole. The various potentials inherent in the knowledge of bead making, herbal medicine, and conservation of the environment among the Maasai people have not been fully explored. These potentials should be harnessed by recentering the voices of the youth and community members, both by the state and private sector operators.

As in other parts of Africa, youth unemployment is high in Kenya. According to the Kenya Federation of Employers (n.d), although the overall unemployment in Kenya is at 12.7 percent, youth (15–34-year-olds), who form 35 percent of the Kenyan population, have the highest unemployment rate of 67 percent. According to the World Bank (2022), youth unemployment among 15-24-year-olds in Kenya is 13.4. When disaggregated on a gender basis, the data show that unemployment among young women is higher at 14.5 percent. To address this problem, the Government, through the Ministry of Industrialization, has put in place various programs to promote small, micro, and medium enterprises in the country. It has also been working with the Kenya Institute of Management, Jomo Kenyatta University of Agriculture and Technology, and Techno Serve, to identify growth-oriented and innovative business ideas that can be nurtured into vibrant and sustainable business enterprises (Wanjori, n.d). Despite the great thoughts that underpin these programs, there is no evidence that a deliberate effort was made to incorporate the indigenous knowledges of the marginalized groups, such as the Maasai and the Ogiek peoples, among others. Although, as mentioned above, the constitution of Kenya has recognized the distinct identity of the Indigenous People in the country, this has not translated into deliberately incorporating their knowledge into the design of employment and entrepreneurship policies.

The (mis) use of state power has affected the extent of engagement with the indigenous knowledges of the Indigenous People in Kenya since colonial times. According to Dominguez and Aydan (2022, 1), for decades, Ogiek have been routinely subjected to arbitrary forced evictions from their ancestral land without consultation or compensation, first by colonial authorities and subsequently by the Kenyan government. The hunter-gatherer nature of their existence has been largely distorted through massive, forced removal from their original ancestral lands in the name of conservation. An engagement with the people to tap into their knowledge of hunting, herbal medicine, and preservation of the environment can help in the co-creation of knowledge required to address the problem of unemployment among the youth of the Ogiek and Maasai people, among others. Although they might not be disposed to living in cities, their natural mode of existence and engagement with nature could still help the government in not only framing appropriate policy responses that can help the people live in more dignified ways, but also in supporting the government in addressing the problem of climate change.

The various interventions have not worked to reduce poverty and inequality or foster the use of entrepreneurship because of the following reasons. First, most of the programs on employment and entrepreneurship in Africa adopt frameworks developed in the Global North, focusing on high-tech innovation, formal markets, and global competitiveness. Where these models can indeed lead to job creation and economic growth, they are disconnected from the live realities of Indigenous people, who are mostly resident in rural areas. More importantly, young people from Indigenous communities lack the education, skill, and wherewithal to function within the economic systems developed from elsewhere. To the extent that these systems do not incorporate their knowledge, respect their culture, and uphold their values, the various Indigenous communities see little or no value in applying those knowledges. These models do not also fit into the pattern of living of Indigenous communities, because some of them are seasonal workers who are constantly on the move. For instance, Maasai young men and women are herders who move from one location to another in search of pastures for their cattle. Any economic program that would be of benefit or even of interest to this group of people would have to be focused on the use of the indigenous knowledge systems that are available within their communities to upscale their activities. Additionally, the foreign model that has been applied in the design of entrepreneurship policies in Africa has ignored indigenous economic practices such as communal farming, cooperative saving groups, and unorthodox apprenticeship systems (Santos, 2014). This point is critical because these indigenous practices have both intrinsic and material values that are not available in the foreign models. For example, the engagement of youths in communal farming leads to social and peer bonding, facilitates shared interests, and enhances optimal use of time. It is more optimal for ten young men or women to work on a farm of their friend in one day than for the friend to do the work alone for two weeks. Even though this is done on a rotational basis, the time saved in the cultivation or harvesting could create additional benefits to the beneficiaries.

Secondly, many of the exogenous entrepreneurship models diverge from the holistic and nonexclusive materialist orientation of indigenous communities. Whereas the whole discourse about entrepreneurship from the exogenous perspective is on the promotion of profit-making, individualism, and the reification of the self, indigenous entrepreneurship is more concerned with communal and ecological values, as well as the consideration for the future (Oloruntoba, 2025). Thirdly, a very poignant failure of the conventional approach to entrepreneurship that has led to their failure in addressing gaps in poverty is the spatial discrimination that is embedded in them. Whereas over 60 percent of Africans across demographic strata live in rural areas (for marginalized communities, this could be 99 percent), there is an urban bias in the design of those policies as efforts are usually concentrated in urban areas (Olaitan and Oloruntoba, 2025). Apart from their spatial exclusion, their voices are also excluded from the design of those policies. The silencing of the voices of indigenous people feeds into the colonial pattern of exclusion and is an indication of coloniality of knowledge in postcolonial Africa (Ndlovu-Gatsheni, 2018). Findings from recent research conducted among the Batwas of Uganda and South Africa, Nguni of South Africa, Ijaw and Ogoni of Nigeria, Maasai of Kenya and Tanzania, show that the youths in these communities have not been consulted in the design of entrepreneurship policies. This lack of consultation is affecting the extent to which the policies have succeeded.[1]

Towards Decolonialization and Recentering the Voices of Indigenous Youths on Employment and Entrepreneurship

The preceding sections of this chapter have shown that the attempts of South Africa, Tanzania, and Kenya to solve the problem of youth unemployment have been informed by the logic of coloniality in which the state has focused on Eurocentric approaches to employment generation and entrepreneurship programs without paying attention to the voices and knowledges of marginalized communities. The decolonization of this process would entail a deliberate acknowledgement of the authenticity of the knowledges of the marginalized peoples as well as their subjectivities. Despite the recent pushback from right-wing political elites in Europe and the United States, as well as their rhetoric on rolling back the gains of the past twenty years, the call for epistemic and policy decolonization has remained steady and insistent.

Scholars of decolonization see it as the process of challenging and replacing dominant Eurocentric epistemologies with plural, indigenous, and context-based ways of knowing (Mignolo, 2009, Grosfoguel 2007). In Africa, which is one of the continents that is worst affected by colonialism and its after-effects, decolonization involves reviving and legitimizing indigenous systems of education, economic organization, and governance that were suppressed during colonialism (Ndlovu-Gatsheni, 2018). It also involves recentering the voices of marginalized youths and community members on issues that concern them. In the context of Indigenous entrepreneurship and access to dignified jobs for the youths, decolonization involves a reengagement with the knowledges and practices of Indigenous communities that have been sidelined, ignored, despised, or erased through colonial and postcolonial experiences of the colonizers and their offspring with black skin (Fanon, 2008).

Decolonization in the sphere of entrepreneurship and job creation is necessary to address the ongoing scourge of poverty and unemployment in Africa. It is also imperative to decolonize so that the economic structures that colonialism established can be reimagined and rooted in the local experiences, knowledge systems, and the worldview of the various communities. Thus, decolonizing youth employment and entrepreneurship strategies entails listening to and learning from elders, traditional leaders, women's associations, and youth themselves in ways that foreground their agencies and subjectivities. One of the ways to sustain the ongoing efforts at decolonization is to be intentional in foregrounding the voices of the youth and members of indigenous communities for whom entrepreneurship and job-creating policies are designed. As Olaitan and Oloruntoba (2023) argue, Africans must now transcend reactions to the epistemological debates and undue silencing of African knowledges by recentering indigenous knowledge in every sphere of life. Recentering indigenous knowledge will not only promote epistemic justice but could also enhance better policy outcomes for the youth in Africa.

Furthermore, recentering the voices of the youth through the co-creation of knowledge in the design of entrepreneurship policies will ensure cultural sensitivity and relevance. Co-creation will ensure that entrepreneurship policies are rooted in the local realities and lived experiences of the people who are the beneficiaries of those policies. As the Yoruba proverbs say, *aaki gerun leyin onirun* (one's hair cannot be barbed in his or her absence), it is also true that to barb or plait hair, one must be available not just to discuss the style of the hair but to afford the hair stylist or the barber to have access to the head. Additionally, recentering indigenous knowledge in the design of entrepreneurship policies will ensure consideration for the environment. As earlier mentioned, while the Western model of entrepreneurship is usually about profit-making, with little or no consideration for the environment, indigenous entrepreneurship focuses on not only profit-making but also the protection of the environment and the future. While innovation is very important for driving job creation, especially in a digital age, incorporating indigenous knowledge will enhance sustainability.

There are various innovations through indigenous knowledge that can be leveraged upon to upscale entrepreneurship activities. Some of these opportunities are found in agroecology and sustainable farming (such as crop rotation, mixed cropping and organic composing; Altieri & Toledo, 2011), traditional medicine and health products (such as herbal skincare and cosmetic brands, community-based wellness centers that use traditional healing, working with scientists for validated phytomedicine production), textiles and fashions (eco-friendly fashion lines using traditional fabrics and dyeing techniques, cultural story telling through wearable art and creation of export platforms for artisan cooperatives). Other innovation opportunities also exist architecture and eco-construction (such as building sustainable tourism lodges with traditional designs, low-cost green housing for rural development, providing expert advice on vernacular design), environmental conservation and climate adaptation (including community forest management, early warning systems and weather forecasting based on indigenous knowledge, eco-tourism rooted in traditional ecological values), cultural education and media (such as building indigenous language apps, creating animated series based on African folklore as well as digital archiving and publishing of traditional knowledge). Other innovation opportunities also exist in water management and natural resources use, food systems, and gastronomy (such as building restaurants or food trucks offering indigenous dishes, packaging snacks from traditional recipes). There are also innovative opportunities for African youth-led businesses in crafts, music, dance, and storytelling that can become sustainable sources of income rooted in cultural heritage (Hoppers, 2002). In the final analysis, it is important to highlight and reiterate that decolonizing knowledge is not about completely rejecting knowledge from other places but ensuring that Africa's search for the intractable challenges of poverty and unemployment is addressed by identifying and utilizing the massive knowledge that is available on the continent. Using the knowledge found in Africa as a foundation for job creation through entrepreneurship will enhance effectiveness and sustainability both in the medium and long term.

Rather than seeing the youths as objects for whom a policy on employment and entrepreneurship should be developed, it is imperative to consider them as equal citizens whose voices must be integrated both in the conceptualization, design, and implementation of such policies. The starting point would be a systematic engagement with the various indigenous communities, their elders, youth, and women in a process of co-creation of knowledge. As mentioned above, African indigenous knowledge systems are authentic and mostly applicable to addressing the social, economic, and political systems in their environments. As Mawere (2015) argues, the African traditional knowledge was used to build the various societies in Africa before the colonial intrusion. Despite the silencing and subordination of these knowledges by the state, they have been resilient (Hountondji, 1997). Their resilience is seen in how various aspects of African knowledges continue to be relevant in our contemporary times. The resilience of these forms of knowledge over the years continues in the indigenous ways of preserving food, rearing animals, preserving the environment, bead making, handicraft, herbal medicine, and the like.

Recentering the voices of the indigenous youth, for instance, would help the government and the private sector to identify areas of priority for their interventions. In this process of co-creation, indigenous youth who have succeeded in specific aspects of economic activities can provide training for upcoming ones. Apart from encouraging peer learning, this would help to preserve the knowledge of the people as well as the values and cultures across generations. All the government would need to do then is to provide all the necessary administrative, technological, and financial support. Listening to the various stakeholders in the indigenous communities in Africa will enhance the quality of policy design and implementation. Knowledge gained from respectfully engaging with the indigenous communities can also be applied to other parts of the country concerned. Rather than limiting the application of this knowledge to the communities, the government and the private sector can leverage it to expand employment and entrepreneurship opportunities, which can benefit the country. For instance, given the large export market for handicraft and bead products (which are mainly produced by members of the Indigenous communities), both the government and the private sector can invest massively in these sectors by equipping indigenous youths and elders to train other youths in the country.

There are policy options for the government to consider on recentering indigenous knowledge and amplifying the voices of the youth and marginalized communities. One of the best policy options is the intentional integration of indigenous knowledge systems into national education curricula and entrepreneurship training at all levels of education, from kindergarten to university level. Including indigenous knowledge and languages in the curricula will foster more learning and inculcate in the learners a sense of cultural identity and pride. It will also ensure that the learners are conversant with the traditions, norms, culture, and ontology of their ancestors and environment (Hoppers, 2002; Wathiong'o, 1981). Since colonial times, teaching pedagogy in Africa has been structured to mimic Eurocentric ideas and cultures. This explains why elementary school teachers use an apple to describe the capital letter A when neither the teacher nor the pupil has ever seen an apple (see Ezeanya, 2016, YouTube).

Technical and vocational training schools should also have indigenous knowledge as part of their curricula. This will allow them to learn and apply some of the indigenous-oriented innovation opportunities that have been identified above. Although centers of entrepreneurship are being established in Africa, not much, if any, is focused on indigenous entrepreneurship. Second, the government must also decentralize program design on entrepreneurship by including rural and traditional institutions. As Olaitan and Oloruntoba (2025) note, there is an urban bias in the design of government policies aimed at poverty eradication and job creation in Africa. Yet 70 percent of Africans still live in rural areas. These spaces are not mere geographical locations. Rather, they are imbued with massive indigenous knowledge systems that can be explored and applied to create dignified jobs.

Beyond the adaptation of existing institutions and knowledge systems for learning, the government should also work to promote cross-generational learning, promoting oral histories, storytelling, and mentorship. Elders are a repository of vast knowledge, which can be passed to future generations. Although retirees and old men and women in communities are passing knowledge to young family members at different levels, the government can support the institutionalization of this process by ensuring that old people in villages and urban areas who might be too old to do formal work are engaged on a contract basis to pass indigenous knowledge to the future generations through oral history and storytelling.

Lastly, the government can support the establishment of community-based enterprises by providing access to land, finance, markets, certification for products, and granting intellectual right protection for indigenous knowledge-based innovations. Unleashing the massive potential inherent in indigenous knowledge requires access to infrastructure such as regular electricity. Dependence on hydrocarbons as a source of electricity has not helped in ensuring energy sufficiency in Africa. Interestingly, there are massive resources in indigenous communities that can be used to tap into opportunities in solar power and renewable energy. The government should also establish indigenous knowledge councils to provide advice based on the knowledge systems of the communities about youth employment strategies.

Conclusion

This chapter has examined the imperative of decolonizing knowledge on employment and entrepreneurship in Africa. The silencing of African indigenous knowledge systems is situated within the politics of knowledge production in which Eurocentrism was universalized as the authentic knowledge, while other knowledges are regarded as inferior. Using the experiences of Kenya, South Africa, and Tanzania in their relations with the Indigenous communities, we argue that there is a colonial present in which the voices of the youth and other community members are silenced and excluded in the design of employment and entrepreneurship policies. It is also noted that despite the coloniality of knowledge that still defines many parts of Africa, Indigenous knowledge continues to show resilience as both marginalized and non-marginalized communities continue to use the knowledge at micro levels. Given the massive potential inherent in these knowledges for upscaling opportunities in job creation and entrepreneurship, we recommend that both the state and the private sector should engage in respectful conversation with indigenous youths, elders, and women in ways that can contribute to harnessing these knowledges for the benefit of both the communities and the African countries at large. We also recommend policy options for the African governments that can lead to recentering the voices of the youth and their communities in the design and implementation of entrepreneurship and employment policies. These policies can help to recenter indigenous knowledge for dignified jobs, innovation, and sustainable development.

Note

1 These findings were part of an ongoing project on Unlocking opportunities in Indigenous knowledge for Entrepreneurship in Africa. The project is undertaken by the Africa Indiegnous Knowledge Research Network and supported by the Mastercard Foundation

References

Adebajo, A., & Whiteman, K. (2010). *The EU and Africa: From Eurafrique to Afro-Europa.* Columbia University Press.

African Development Bank. (2016). *Jobs for youth in Africa: Strategy for creating 25 million jobs and equipping 50 million youth.* African Development Bank.

Ake, C. (1981). *A political economy of Africa.* Longman.

Altieri, M. A., & Toledo, V. M. (2011). The agroecological revolution in Latin America: Rescuing nature, ensuring food sovereignty and empowering peasants. *Journal of Peasant Studies, 38*(3), 587–612.

CABAN Investments. (n.d.). *Promoting entrepreneurship in South Africa: Lessons from successful national start-up ecosystems.*

Colbourne, R. (2017). Indigenous entrepreneurship and hybrid ventures. In A. Corbett & J. Katz (Eds.), *Perspectives and approaches to blended value entrepreneurship: Advances in entrepreneurship, firm emergence and growth* (Vol. 19, pp. 1–?). Emerald Publishing.

Colbourne, R., Peredo, A. M., & Henriques, I. (2023). Indigenous entrepreneurship? Setting the record straight. *Business History*.

Conrad, J. (1990). *Heart of darkness*. Dover Publications.

Di Giorgio, A. V., & Habibis, D. (2019). Governing pluralistic liberal democratic societies and metis knowledge: The problem of Indigenous unemployment. *Journal of Sociology, 55*(1), 37–53.

Dominguez, L., & Aydan, F. (2022). *Reparation at last: Land justice for Kenya's Ogieks*. Minority Rights Group International.

Ezeanya, C. (2016). *How Africa can use its traditional knowledge to make progress* [Video]. TED. https://www.youtube.com/watch?v=28sa2zGgmwE

Easterly, W. (2006). *The white man's burden: Why the West's efforts to aid the rest have done so much ill and so little good*. Penguin Random House.

Fanon, F. (2008). *Black skin, white masks*. Grove Press.

Federation of Kenya Employers. (n.d.). *Youth employment*. https://www.fke-kenya.org/policy-issues/youth-employment

Goats and Soda. (2014, December 22). *The Khoisan once were kings of the planet. What happened?*

Global Citizen. (2021, August 10). *South Africa's Indigenous people in a land dispute against Amazon over new Africa headquarters*.

Grosfoguel, R. (2007). The epistemic decolonial turn: Beyond political-economy paradigms. *Cultural Studies, 21*(2–3), 211–223. https://doi.org/10.1080/09502380601162514

Hountondji, P. J. (1996). *African philosophy: Myth and reality* (2nd ed.). Indiana University Press.

Hountondji, P. J. (1997). *Endogenous knowledge: Research trails*. Council for the Development of Social Science Research in Africa.

Ighobor, K. (2017). Africa's jobless youth cast a shadow over economic growth. *Africa Renewal*.

International Working Group on Indigenous Affairs. (2019). *Tanzania*. International Working Group on Indigenous Affairs.

International Work Group for Indigenous Affairs. (2024). *Indigenous peoples of South Africa*. https://iwgia.org/en/south-africa/5358-iw-2024-southafrica.html

Nyiagaraa, K. (2022, November 27–December 2). Indigenous youth and entrepreneurship in Ogoni, Nigeria [Conference presentation]. Africa Studies Association Roundtable on Indigenous Youth and Entrepreneurship in Africa, Philadelphia, PA, United States.

Mawere, M. (2015). Rethinking African indigenous knowledge and world civilizations: Pasts, presents, and futures. In M. Mawere & S. Awuah-Nyamekye (Eds.), *Between rhetoric and reality: The state and use of indigenous knowledge in post-colonial Africa* (pp. 1–18). Langaa RPCIG.

Memmi, A. (1991). *The colonizer and the colonized*. Beacon Press.

Mignolo, W. D. (2009). Epistemic disobedience, independent thought and decolonial freedom. *Theory, Culture & Society, 26*(7–8), 159–181.

Mwakalila, E. (2022). Income inequality: A recipe for youth unemployment in Africa. *SN Business & Economics, 3*(1), Article 15. https://doi.org/10.1007/s43546-022-00394-0

National Economic Empowerment Council. (2017). *Tanzania inclusive national entrepreneurship strategy*. Government of Tanzania.

Ndlovu-Gatsheni, S. J. (2018). *Epistemic freedom in Africa: Deprovincialization and decolonization*. Routledge.

Nsiku, D., & Henry, C. (2021). Policies for supporting women entrepreneurs in developing countries: The case of Tanzania. *Journal of Entrepreneurship and Public Policy*.

Santos, B. de S. (2014). *Epistemologies of the South: Justice against epistemicide*. Routledge.

Ochonu, M. E. (2015). The Wangara trading network in precolonial West Africa: An early example of Africans investing in Africa. In T. McNamee, M. Pearson, & W. Boer (Eds.), *Africans investing in Africa* (pp. 21–35). Palgrave Macmillan. https://doi.org/10.1057/9781137542809_2

Okolo, P. (2022, November 27–December 2). Youth perspectives on entrepreneurship and employment in indigenous communities among the Ijaws in Nigeria [Conference presentation]. Africa Studies Association Roundtable on Indigenous Youth and Entrepreneurship in Africa, Philadelphia, PA, United States.

Oloruntoba, S. O. (2025). *Decolonizing knowledge on youth employment and entrepreneurship in Africa: Why the voices of Indigenous youth matter?* Bloomsbury.

Olaitan, Z. M., & Oloruntoba, S. O. (2025). *African indigenous knowledge systems and the African start-up ecosystem: Partnership for impact*. AIKRN Policy Brief.

Oloruntoba, S. O., & Akinboye, S. (2025, June 3–6). Reframing the grammar of economics: Indigenous knowledge and economic transformation in Africa [Conference presentation]. Canadian Association of African Studies Annual Conference, University of Toronto, Toronto, ON, Canada.

Oloruntoba, S. O. (2023, May 27–June 1). The state and indigenous knowledge in Africa: Towards recentering lost voices [Conference presentation]. Canadian Association of African Studies Conference, York University, Toronto, ON, Canada.

Oloruntoba, S. O. (2014). Social sciences as dependency: State apathy and the crisis of knowledge production in Nigerian universities. *Social Dynamics, 40*(2), 338–352.

Rodney, W. (2018). *How Europe underdeveloped Africa*. Verso Books.

Tshikovhi, N. (2021). A theory of indigenous entrepreneurship in the modern African state: The case of South Africa and Zimbabwe. *International Journal of African Renaissance Studies, 16*(2), 21–60.

Wanjori, A. (n.d.). *SME policy in Kenya: Issues and efforts in progress*.

wa Thiong'o, N. (1981). *Decolonising the mind: The politics of language in African literature*. James Currey.

World Bank. (2022). *Unemployment, youth total (% of total labor force ages 15–24) (modeled ILO estimate): Kenya*. World Bank.

Zeleza, P. T. (1993). *A modern economic history of Africa*. Council for the Development of Social Science Research in Africa.

Chapter 3

RETHINKING THE LANGUAGE OF ECONOMIC POLICY IN AFRICA: INFORMALITY, INDIGENOUS ENTREPRENEURSHIP AND EPISTEMIC TRANSFORMATION

Introduction

The economic structure in most of the countries in Africa is rooted in the historical and contemporary dysfunctionality brought about by colonialism, neoliberal global capitalism, and an incipient Indigenous economic system, which has been generally defined as the informal sector (Ake, 1981). As Ake argues, the colonial economy altered the economic systems in ways that forced a sudden change from a subsistence economic system based on agriculture and trading networks into a wage economic system. This extraverted economy was rooted in Western epistemology, which delegitimizes and subordinates Africa's knowledge systems (Oloruntoba, 2025). The insertion of Africa into the circuit of the global capitalist economy created conditions that still perpetuate economic structures that otherwise rather than integrate the indigenous economic activities. For the convenience of continuity of neocolonial praxis of epistemological and policy framing, a dichotomy has subsisted along the lines of formal and informal economy. While the so-called formal economy receives all the support and patronage from the government, the informal sector faces neglect, abuse, and extortion from state officials (Oloruntoba, 2018). Notwithstanding, it has shown strong resilience as it continues to be the sector that creates jobs for over 80 percent of the population in many countries (ILO, 2018). Despite its scale and resilience, the informal economy is often framed in policy discourse as a problem to be solved rather than a system to be understood (Ohnsorge and Yu, 2022)

The informal economy in Africa is said to be large, unregulated, and exists outside of the formal economy (ILO, 2018, 2002). Another side of this coin notes that it is linked to the formal economy and integral to development

(Asmal et al. 2024). Informal economy cuts across several sectors, such as street vending, transport systems, and waste recycling in urban areas, smallholder agriculture, artisanal mining, agro-processing in rural areas, as well as informal trade networks linking African markets through cross-border trade. They can also be divided into three types, namely survivalist (subsistence trading, petty services), micro-enterprises such as small-scale manufacturing, and retail, as well as growth-oriented informal firms made up of very dynamic but unregistered businesses. Yet as Mhando and Kiggundu (2018) argue, informality in Africa is quite complex and requires deeper and multidimensional approaches to research. Although the informal economy is often associated with illicit, illegal, and criminal-oriented economic activities, these authors argue that these labels are not limited to the informal sector. They note rather succinctly that 'incidents of illegality, illegitimacy, criminality, low productivity, tax evasion, illicit flows, and other harmful conduct to society exist both in formal and informal economies, perhaps even more so in the former' (p. 219).

The different understanding of what constitutes the informal economy has produced studies that seek to understand its operations, uniqueness, and most importantly, how to formalize it so it can be formal. The conceptual starting point of these studies is the necessity to transform the informal economy and pave the way for a much more formal and developed economy. Attempts have also been made to rationalize why the informal economy exists, with poverty, unemployment, underdevelopment of African economies, etc., being the most echoed (Oloruntoba, 2018). This posits that the informal economy exists due to the shortcomings of most African states and the stunted growth that African economies find themselves in, a rationalization that ignores the uniqueness of the indigenous economy and is not just a product of developmental pitfalls but the contextual specificity of most African societies. Self-employment, sole-proprietorship, trading, home-based entrepreneurship, etc., are integral to African societies.

A quick trace of Africa before colonization proves the existence of these types of economic activities, which are now part of Africa post-colonization (Akyeampong 2017). How then do these activities get termed as informal, such that they are seen as consequences of development failures rather than as relevant economic activities that form part of the economy? An answer can be found in Francis Fukuyama's declaration of liberalism as the dominant ideology and Western capitalism as a universal economic system (Fukuyama 1992). Or the fact that Bretton Woods institutions adopt a Western approach to development and economic growth. This can also be traced to the obsession with GDP as a measure for growth and development, activities that cannot be captured within this are outliers and therefore unacceptable

(Oloruntoba, 2016; Fioramonti, 2014). The International Labor Organization is leading the attempt to formalize the vast informal economy in most African countries to smooth the development path (Kiaga and Leung 2020). This attempt is aimed at ensuring that informal workers are integrated into the formal economy. As with many other external interventions and attempts at providing solutions to the challenges of poverty and employment in Africa, this shows a lack of understanding of history, contexts, and the reality of how Africans have organized themselves in terms of economic activities before the colonial intrusion.

In this chapter, we engage the nuance around the categorization of economies into formal and informal to highlight this problem. It critiques the very essence of the ILO, dividing economies into formal and informal, such that the informal has to shed itself to become formal. We examine the concept of informality and how it has come to be used to describe most African economies. It argues that to rethink the language of economic policy, we must engage the foundation of dominant economic models and paradigms that recognize Western theories as the acceptable pathway, while alternative models are seen as outliers. This chapter posits that framing economic activities that contribute significantly to the Gross Domestic Product and create jobs for more than 50 percent of the population in some instances, are rooted in colonial epistemologies that misrecognize the structural and productive role of African indigenous economic systems. Drawing on political economy, postcolonial and decolonial theory, and comparative case studies of Nigeria, Kenya, South Africa, and Rwanda, we advance a critique of the language of informality and propose pathways for transforming what is generally regarded as informal activities into viable, inclusive, and sustainable economic systems. The mischaracterization of this important sector has resulted in abuse, extortion, and limited support from the government for the operators in the informal sector, both at the national level and cross-border exchanges (Oloruntoba, 2018; Nshimbi, Moyo, and Oloruntoba, 2018). Thus, it is imperative to rethink the language of economic policy in Africa. This necessarily means that rethinking the language of economic policy must be preceded by a deliberate de-centering of Western economic models and a recognition of Indigenous entrepreneurship (IE) as a legitimate and generative form of economic practice. Such an approach enables the contextualization of development, expands the analytical lens through which African economies are understood, and repositions Indigenous entrepreneurial practices as central rather than peripheral to economic policy and transformation.

The chapter is divided into five sections; the first section traces the origin of informality to set the background for how it came to be used in economic policy in Africa. The section examines what informality is, the informal economy, and the different understandings of the concept. The third section examines the prevalence of the informal economy in Africa. The fourth section provides a critique of Western economic models while showing how the ILO dichotomization of economies is a manifestation of these models. The final section fits the preceding sections together while establishing the need to rethink the language of economic policy in decolonial ways that also recognize IE.

Tracing the Origin of Informality

In tracing the origin of the dichotomization of economic practices in contemporary Africa, it is important to locate this within the colonial enterprise, which fundamentally altered the economic structure of the continent (Rodney, 1972). By focusing on extraction, exports, cash crop production, and wage economy, the colonial system created a dichotomy that resulted in formal and informal economic structures. The formal sector was focused on satisfying the interests of the colonial governments through the export of raw materials and imports of manufactured products, while the informal sector comprised indigenous economic practices. This dichotomy had fundamental implications for access to capital, institutional support, and patronage. As Skar (1966) argues in the case of Nigeria, the British colonial system adopted a preferential system in which only white-controlled and, to some extent, Lebanese-controlled companies had access to the required capital to do business.

More poignantly, the colonial systems created a wage economy that forced Africans to leave the economic activities to which they were naturally accustomed to new ones created by the colonialists. For instance, the colonialists introduced new products like cocoa, coffee, tobacco, rubber, and tea, among others. As Ake (1980) argues, the introduction of these products led to a dramatic shift in the economy of the colonized societies. These products were introduced to serve the industrial bases of the colonialists while replacing the subsistence economic activities of Africans. To generate revenue for the metropolitan government, Africans were also forced to pay taxes, with defaulters being subjected to heavy fines, including imprisonment.

Research on informality is founded on popular economic theories/models in the traditional sector throughout the 1950s. Economists like Arthur Lewis thought that as economies expanded, the enormous pool of excess labour in developing countries, including small-scale, casual wage workers, and petty traders, would progressively be absorbed into the formal industrial sector (Lewis 1954). Economic anthropologist Keith Hart contributed to the emergence of attention on informality based on the study he conducted in

Ghana, where he noted that the informal sector had not only survived but grown in the early 1970s (Hart 1973). In addition, the ILO commissioned a team of experts in 1972 to examine economic activities in Kenya. Both Hart's study and the ILO mission in Kenya led to the adoption of the term informal sector. The ILO 1972 mission found that economic activities that make up the informal sector were largely unrecognized, unrecorded, unprotected, and unregulated (ILO 1972).

They noted that various activities, from marginal survivalist labour to lucrative businesses, were included in the informal sector (ILO 2010). The 2002 International Labour Conference Resolution adopted the term informal economy in place of informal sector to properly capture the vastness and complexity of this global phenomenon. The resolution offered guidelines for understanding the informal economy and the different characteristics of this economy. Informality has been seen to be distinct from formality, giving base to the narrative that the informal economy is marginal and unrelated to either the formal economy or the rise of contemporary capitalism (ILO 2009).

Since the concept of informality was first developed in the 1970s and its dimensions were first understood through narrow terminology such as the informal sector, the ILO has been at the forefront of the analysis of this phenomenon. However, opinions in development circles regarding the informal sector as a reality of the economy were divided. The informal sector was viewed by many as marginal or peripheral, unrelated to either the formal sector or the advancement of contemporary capitalism. Some of these observers thought that once emerging countries like Ghana, Kenya, and others reached a certain point in their contemporary industrial development or economic growth, the informal sector would vanish. Some contended that the trajectory of industrial development in developing countries may diverge from that of rich countries, encompassing the growth of informal economic activity (Chen 2012).

In many economies and countries, the 1990s saw an increase in informality due to economic globalization (Standing 1999). Globalization created new opportunities and opened doors to new markets, but many of these jobs were not 'formal' jobs; they were small-scale or underprivileged producers who could not access many of the new markets. The increased attention was also because the informal economy expanded globally and manifested in unexpected areas. During the economic recession in the early 2000s, informal employment increased dramatically (Horn 2009). The understanding of the connections between informality, poverty, and inequality, as well as the connections between informality and growth, is another factor contributing to this resurgence of interest. More people are realizing that a large portion of the contemporary informal economy is interdependent with the formal economy and boosts the entire economy. They also understand that helping the working poor in the informal economy is a crucial step in the fight against poverty and inequality. Furthermore, it is becoming more widely acknowledged that women are disproportionately employed in precarious informal sectors, making it imperative to assist impoverished working women in this informal sector as a means of eliminating both gender inequality and poverty among women (Chen et al. 2005). Keith Hart had a positive outlook on the informal economy; he noted that the majority of internal migrants in Accra were involved in informal economic activities with an autonomous capacity for generating income, although they were subjected to external restrictions and capitalist dominance (Hart 1973). Also, the Kenya Mission emphasized how unorganized the informal sector was and its potential for reducing poverty and generating jobs (ILO 1972).

Understanding the Informal Economy

The ILO 2002 Resolution provided a clear conceptualization of the informal economy. They defined the informal economy as "all economic activities by workers and economic units that are – in law or in practice – not covered or insufficiently covered by formal arrangements" (ILO 2002). Olaitan (2024) explains that the definition covers a broad spectrum of employment and economic activities that lack work-based safety nets, such as street vending, home-based employment, waste collection, domestic work, and temporary contract work, etc. This new framework for comprehending the informal sector was made available at the 2002 International Labour Conference (ILC), which also distinguished between formal and informal economies. ILO (2009) notes that the classification of the informal economy aims to facilitate the transition and integration into the formal economy. Another feature of this classification is the idea of exclusion, which refers to working people who are primarily shut out of the economic activities that occur inside the formal system. This presupposes that workers in the informal economy are overlooked in the creation of policy because they are not included in official statistics and national accounts (ILO 2010).

The lack of social safety nets, the inability to exercise rights at work, and the absence of representation and voice in the workplace are some of the factors that characterize workers in the informal economy (ILO 2002). Hence, the conclusion is that the informal economy is primarily unprotected, unstable, and susceptible because of its exclusion from the advantages offered by the government, private markets, and political processes (Medina et al. 2017). Feige (2005) adds that the informal economy is made up of businesses that eschew legal requirements and are not entitled to the protections and advantages found in laws and regulations pertaining to financial credit, social systems, labor contracts, property connections, commercial licenses, and torts. The inability to measure informality since workers operating in it often have little to no access to employment benefits or social safety, and their working practices erode labour market inclusivity, informs his argument. For developing countries that require a solid tax base, informal economic activity significantly reduces tax income. As a result, emerging countries have the motivation to comprehend the extent of informal economic activity and figure out ways to transition from informality to formality.

The ILO categorization of the informal economy provides a clear picture of the different challenges that workers within the economy face, in addition to a lack of registration and recognition (ILO 2002). The informal economy was historically perceived as a refuge for individuals who were unable to get jobs in the formal economy and were thus compelled to accept any job opportunity or generate their own employment through modest, even peripheral, economic pursuits. However, not all members of the informal economy are impoverished and find themselves in helpless situations. Some began their careers on the margins, yet their dynamism allowed them to expand their businesses and escape poverty (ILO 2010). Additionally, some formal businesses may opt to maintain their informal status in other areas, such as by not filing tax returns or registering their owners and employees with the labor ministry or social security system. Nevertheless, the majority of people working in the informal economy are still impoverished, have low salaries, and have little autonomy concerning where they work (ILO 2002). Even in cases where members of the informal economy can make more money than their formal counterparts, they are still excluded from the primary channels of governmental protection systems and market exchange in several ways.

One of the main arguments concerning the nature of workers in the informal economy is whether, contrary to what the ILO has frequently claimed, informal work is primarily the result of poverty or whether, as De Soto (1989) and Maloney (2004) have argued for Latin America, the informal economy is made up of entrepreneurs who choose to operate there rather than having to deal with the onerous regulations that small business owners must comply with.

Given that the advantages of being informal outweigh the disadvantages of formal employment, workers who choose to be informal do so voluntarily. This analysis takes into account financial factors like income and taxes, but it also considers other benefits of informality, like independence, labour flexibility, and being one's own boss. This analysis best describes voluntary informality as different from induced. Induced informality refers to those informal workers who are forced into informal employment due to overly stringent labour laws or unspoken social norms, such as labour discrimination, despite their willingness to work formally and their productivity, which permits them to be employed freely in this market (Chen 2012). It is thus safe to highlight that informal workers cannot be viewed as a single, homogenous group and that informality has costs to society that should not be disregarded, especially in situations where it may foster inclusive growth for vulnerable populations.

Part of the complexity of workers in the informal economy is the notion that wage jobs are formal; however, some wage jobs are more likely to be informal than others due to unclear employment relationships. For instance, individuals who are employed as paid household domestic workers by formal or informal businesses without making social security contributions are involved in informal wage employment. Informal wage employment consists of workers in unregistered or undeclared businesses, contract workers, casual or day laborers, temporary or part-time workers, paid domestic helpers, employees of informal businesses, and industrial outworkers, sometimes known as homeworkers (Chen 2012). With the addition of employment ties that are neither socially nor legally protected, this extended definition broadens the focus of businesses that are not subject to legal regulations. It also helps to draw attention to informal laborers, or people who work for pay.

The focus on employment has resulted in a substantial reevaluation of the nature, origins, and effects of informal work. These days, it is commonly acknowledged that informal employment encompasses a variety of self-employed individuals, primarily employed in unregistered or tiny unincorporated businesses, as well as a variety of wage workers who are engaged in employment without their employer making any contributions to social protection (ILO 2010).

Self-employed workers have traditionally been excluded from the definition of formal employment. Due to their employment being obscured, unclear, or undefined, a growing number of wage workers have discovered that they are effectively unprotected under the law (ILO 2002). By making it seem like a relationship with a distinct legal status, the employment relationship is purposefully concealed. For instance, a lead firm in a subcontracting chain can assert that it has a commercial relationship with those individuals, instead of claiming to have a subcontracted employment relationship with people who create items for it. There is uncertainty regarding whether an employment relationship actually exists because it is objectively ambiguous. This is the situation, for example, with street vendors who sell things on commission for a distributor or who rely on a single supplier for supplies. There is no doubt that there is an employment relationship, but it is unclear who the employer is, what rights the employee has, and who is in charge of protecting those rights.

In the context of value chain production, for instance, it is unclear if the lead company, the supplier company, or the subcontractor is the true employer. Similarly, when it comes to temporary labour, it is unclear who the true employer is: the companies hiring the workers temporarily or the agency providing them (Chen 2012). Workers in each of these employment arrangements typically lack collective bargaining agreement coverage and labour law protections; in other words, they are considered to be informally employed. It is important to note that, in many such cases, the employer seeks to disguise the employment relationship or avoid the definition of who is responsible, and that the employer in question may well represent a formal firm, not an informal enterprise. It is crucial to remember that, in many of these situations, the employer tries to hide the nature of the employment relationship or avoid defining who is in charge. Additionally, the employer in question can very well be a formal company rather than a non-formal venture.

Informality and the informal economy are seen as a factor of underdevelopment, which creates a negative outlook for it. IMF (2017) sees the informal economy as a vital safety net that gives work and income to a significant number of individuals who could otherwise be forced into poverty when there are insufficient possibilities in the formal sector. This perspective supports the legalist and induced lens that informality exists because of the insufficiency of formality, rather than acknowledging that individuals can choose to participate in the informal economy. Another reason for the negative outlook that informality has is the inability to measure it. Nshimbi (2018) points out that the term informality, more especially, the informal economy, is contested. There is no precise statistical aggregate that can be used to quantify the informal sector because it is difficult to scientifically analyze and list connected activities. Therefore, it is challenging to have a thorough, comprehensive grasp of its entire economic scope and the effects of activities inside it.

Theorizing Informality

Based on the discussions in the previous sections, there are two distinct schools of thought when theorizing informality and the informal economy. The first viewpoint pits structuralists against dualists, while the second pits legalists against voluntary views. The structuralist perspective notes that there is an integral link between the formal and informal economy, and they are components of a single, integrated economic system. This perspective posits that the informal economy, which is made up of economic entities and laborers who work for and are subject to big formal enterprises, has existed and grown because of the nature of capitalist development (ILO 2010). By providing them with cost-friendly goods and services, the informal economy increases the competitiveness of formal sector businesses through this relationship. Because it helps formal sector businesses save labor and consumer costs, the informal economy promotes accumulation in this way. This position is made even though there are claims that the formal economy has become more informal due to layoffs that drive workers into the informal sector. On the other hand, the dualist perspective contends that the informal economy is separate from and unrelated to the formal economy (de Soto 1989). The dualist perspective has its roots in the writings of Arthur Lewis. By making a distinction between the industrial sector, which is located in metropolitan areas, and the "subsistence" sector, which is located in rural areas, Lewis's two-sector model alludes to this dualism (Lewis 1954). There are further theoretical attempts to comprehend informality from the perspectives of voluntarism and legalism. According to the legalist perspective, the strict control of economic activity in the formal sector is the root cause of informality. Governments and mercantilists (or businesses) work together to create cumbersome regulations that make the legal system unfriendly to those operating in the unorganized sector. Therefore, informality offers a setting where independent contractors can save time and money by evading the onerous formalities of formalizing or registering their businesses. According to the voluntarist perspective, informal business owners consciously leave the highly controlled formal economy. They act in this way to elude official laws and taxes (Nshimbi 2018).

ILO Dichotomization as a Consequence of the Western Economic Model

The ILO classification of the informal economy and the concept of informality broadly is founded upon Western economic models/theories of development. The idea is that development must manifest in a certain way, and anything that does not conform to that is not recognized as development. Previous studies, such as Alvarez et al. (2011), attempting to understand Africa's development trajectory have taken the approach of either employing crisis narratives based on preconceived notions of neopatrimonialism or inherent African cultural primitivity, or doing trend analysis of economic development in Africa using traditional, established indices like GDP (Njoh and Ayuk-Etang, 2012). Renowned African economists like Augustine Fosu have had success "getting the numbers right" and demonstrating that Africa's GDP performance is actually rather good (Fosu, 2010). These earlier attempts, however, fall short of providing a comprehensive analysis of the skewed framing of development that has been universally adopted, including its definition, methods of measurement, and practices. They also fail to address the reasons behind development's pursuit, specific methods of measurement, and current state, all of which are central to the debate over whether economic development in Africa is flawed (Bierschenk and Spies, 2010).

Since 1945, a process of change has been the accepted definition of economic development. Nonetheless, this shift has always only applied to government initiatives to alter the characteristics of things like land tenure (Obeng-Odoom 2013). Although economic development was viewed as a necessary process of social change, the results of that change experienced significant alteration. (Obeng-Odoom, 2013). Economic development in the past has been associated with modernization, industrialization, and even Westernization (Arndt, 1989:2). With the end of World War II and the independence of the former colonies, which came to be known as the Third World, economic growth gained international recognition as a goal of economic development.

Like Adam Smith, most economists believed that the primary goal of economic progress should be capital accumulation, particularly in the context of industrialization. This assumption is based on the theory of the big push, which gained a lot of traction and maintained that developing countries needed a significant infusion of capital to give them a "big push" because they were trapped in a specific state of widespread unemployment and underemployment in the agricultural sector. Bauer condemned their writings, claiming that they were at odds with actual data. He maintained that there was virtually no external assistance in the path of the then-poor in the process of being developed, that some poor countries had achieved significant progress, and that the currently rich countries had once been impoverished but had escaped it (Bauer 1971:34).

Others, like Albert Hirchman, identified other growth-promoting factors, chief among them the necessity of fostering entrepreneurial abilities in less developed nations (Martinussen, 2005:59). On the other hand, Lewis presented one of the most significant theories in the field of development economics about the growth of the private sector, employment, and economic development. Lewis (1954) coined the term dual economy model to describe the process of economic development with unlimited supplies of labor. Lewis maintained that two key features of economically underdeveloped countries were unrestricted labor and a bifurcated economy (one for sustenance, the other for capitalist use). He claims that as the economy develops, a sizable subsistence sector like agriculture loses surplus labor to a comparatively small capitalist sector. Therefore, an increase in capitalist profits is good for economic growth because they reinvest what Lewis (1954) called the 'capitalist surplus,' which spurs capital accumulation and moves surplus labor from the economy's subsistence sector into the modern sector. While other forces for change were acknowledged, the ultimate goal was usually economic gain. In other words, economic development was equated with economic growth as an end in and of itself. Yet as Seers (1979) would argue, economic growth is not the same as development, as there can be growth without development. This argument has been proved to be true in several African countries where growth in gross domestic product has not necessarily led to improvement in the living conditions of people (see Oloruntoba, 2016).

Economists discovered in 1965 that while Third World economies were growing at times even faster than First World ones, issues like inequality, sickness, illiteracy, and substandard housing persisted. Concerns about the wider social objectives of economic development began to replace the growth-centered perspective on the subject. Easterly (2009) has persistently attempted to demonstrate that the MDGs are anti-African. He demonstrates how, given their existing low base, African countries are likely to be more favorably impacted by UN assessments of progress or regress than other countries. Additionally, 1990 serves as the foundation year for analysis, even if the MDGs were widely adopted in 2000. It should be noted that during the 1990s, the majority of African nations faced challenging economic conditions. As a result, from the outset of putting the MDGs and associated methods into practice, Africa received a failing grade.

Mainstream economic models reinterpreted the concept of development to mean modernization, which they viewed as a universal singular path of progressive change, to inform US military intervention and economic assistance in post-colonial areas (Gilman 2003). A smug celebration of market democracy's ability to bring about development grew to rule the world of Western aid. The policies coming out of most international development agencies and developing countries mirrored this neo-liberal Washington Consensus, which was primarily focused on growth-oriented initiatives at the expense of social issues (Mahon 2017). Western institutions and specialists were obsessed with modernizing Africa and the developing world, and in order to do so, they depended on Western economic models/ theoretical tenets based on modernization theory. According to this view, for traditional or developing countries to advance, they must take the same paths as developed nations in the West (Rostow, 1961). According to modernization theorists, all civilizations are evolving from traditional to modern through a series of stages in an evolutionary process (Noyoo 2015).

The above explanation underpins the assertion that informality is seen as a feature of Africa's underdevelopment and attempts to turn informality into formality are an offshoot of the modernization theory that argues for developing countries to tow the same path as developed countries. The ILO classification is based on a certain idea of what development is and what should drive it. Therefore, the inability of informality to be accommodated in the modernization-based idea of development marked it as an outlier and a challenge. Models used to understand Africa or developing countries are devoid of the social reality and context of these regions. Noyoo (2015) asserts that theories and paradigms that originate from a given social context will reflect factors and circumstances unique to that setting, such as the environment. Because they are drawn from a context significantly distinct from Africa, they are unable to adequately address special issues that arise when applied to the interpretation of conditions in other countries. African countries are still receiving development guidelines from the West that are based on their historical and sociopolitical circumstances. As a result, the development paradigms that experts rely on to identify potential solutions for African nations are typically paradoxical (Noyoo 2022).

Africa and the Informal Economy

De Soto (1989) establishes that the informal economy is the people's unexpected and innovative solution to the government's inability to provide the basic requirements of the lower class, especially through wage work that would enable them to support themselves. Africa is one of the continents with the largest informal economy, with most of its population working in the informal economy rather than the formal economy (Kiaga and Leung 2020). Statistics show that the informal sector accounts for 60–85 percent of employment generation in several African countries and contributes over 42 percent to the Gross Domestic Product in some (ECA, 2014). Data from the ECA further shows that the scale of informality has barely shifted over the past two decades, with the share of informal employment declining only marginally from 84.3% in 2005 to roughly 83% in 2024 (ECA, 2014, The North Africa Post, April 18, 2026).

A brief tracing shows that informality and the informal economy have always been characteristics of the African economic system. During the colonial era, the formal/informal dichotomy denoted variations in how colonial power was recognized. However, during the postcolonial and independence eras, it changed to represent access to social services, money, and means of subsistence. This implicitly reflected the supply and demand dynamics of the market, whereas the former marked the beginning of the formal/informal sector's legal recognition (ibid). In ignorance of history, it is argued that the expansion of the informal economy in Africa is a function of the developmental failures that it has and is going through.

The 1980s were often referred to as Africa's lost decade, during which time globalization began to flourish, and developmentalism began to falter. During this time, several African countries embraced the Washington Consensus economic models of development, which brought about free market economies, strict monetary policy to control inflation, and a limited role for the government that resulted in its removal from many of its previous undertakings (Obeng-Odoom 2013). This era and the subsequent Structural Adjustment Programs (SAPs) have been identified by numerous academics using the developmental approach to informality as the primary causes of informality in Africa (Mkandawire and Soludo, 1999). Due to these developments, the continent saw a rise in worker differentiation and income inequality, a weakening of labor market contracts and informal employment, a gender gap in socially assigned roles within the home and society, and an increase in labor market flexibilization (Fenwick et al. 2007). This argument establishes that most people in Africa join the informal economy out of a lack of opportunity in the formal sector and a lack of other sources of income, rather than voluntarily. Their arguments have been based on the rationale that the current policy climate is favorable to free trade, financial liberalization, market deregulation, privatization, labor market flexibility, production outsourcing, and the rise in temporary and formal employment, so participation in informality is not out of choice. These arguments are also informing the attempt to transition from informality to formality in Africa. Africa's policy agenda has made the shift to informality a priority due to the significant shortage of decent jobs that is linked to the informal sector, as well as its detrimental effects on equitable and sustainable development.

ILO (2017) projects that Africa's high rate of informality will continue to be a problem, requiring more formal employment creation to accommodate the growing population, which would mean a shrinking of the informal economy. By 2030, it is expected that 105 million more people will be in Africa than in 2030, with 94 million of them residing in the sub-Saharan region (ILO 2017). While there is a regional tendency towards an increase in informal work, certain countries have managed to lower their percentage of informal employment during the past few decades. However, it did not lead to a reduction in the rate of unemployment and poverty because the formal sector can only absorb a small proportion of the population due to its small size and limited infrastructure for expansion.

While informality in Africa is seen as a problem, scholars have explored the potential that it has for aiding Africa's development. Nshimbi (2018) examines the potential that the informal economy has for integrating trade on the continent. He notes that in the case of informal trade, uncaptured trade that takes place across the vast, porous borders that divide the various African Union (AU) member states' territories underreports the volume of trade in Africa. However, the informal economy and its associated activities form an important and realistic way to integrate Africa, both from the bottom up and from the periphery of nation-states' boundaries. Regarding borders, this would result in borderlands becoming amorphous areas where areas that were once clearly demarcated, that is, which bound and marked distinct communities and polities, citizens, economies, and cultures into disparate sovereign entities, are dissolved by the migrating, trading, and socioeconomic interactions of grassroots actors across the demarcation lines. As a result, integrating African economies from the margins and from the bottom up can be achieved in large part through the informal economy. This is especially true when one considers boundaries and the areas that border them. It is possible to change the informal economy from discrete communities, polities, economies, cultures, and diverse citizenships with distinct borders between nation-states to hazy geographic areas (ibid). Asmal et al. (2024) explain that the informal economy in Africa has the potential to serve as a point of first entry into the labour market for many unemployed people considering the formal economy is too small and not expanding quickly enough to accommodate the number of individuals in need of employment, More specifically, poorer countries tend to hire a large number of people as own-account, self-employed workers, as multiple academics have shown (La Porta & Schleifer, 2014; Poschke, 2019). Self-employed individuals who do not employ anyone outside of their immediate family are known as own-account workers. In fact, more than 70% of people of working age in developing countries may be hired as own-account employees.

The informal sector in Africa has faced several challenges, owing largely to its mischaracterization and the use of Western framings. Some of these challenges include structural constraints, regulatory barriers, lack of social protection, and challenges with productivity (Masina and Mgoduka, 2026). The structural constraints include limited access to credit and finance, poor infrastructure, and restricted market access. By framing the virile African Indigenous economy as informal, the banking sector considers it too risky for access to credit. Access to credit is contingent on providing collateral security, such as land. By the nature of African society, land is usually owned on a communal basis, thus making it possible for individuals to have titles to land. This is especially dire for women, many of whom, on account of gender discrimination, find it difficult to own land.

Another structural constraint facing the informal economy is a lack of quality infrastructure, such as electricity, good road networks, and digital connectivity. For instance, access to electricity is severely limited, especially in rural areas. In a country like Nigeria, households as well as small and micro-enterprises are forced to own generators through which they can generate power. The cost of maintaining these generators represents a huge drain on the limited capital of operators in the informal economy. The informal economy is also affected by restricted market access. Difficulty in moving products from the hinterland due to a lack of good transportation networks and storage facilities affects the extent to which products can be moved to urban areas where markets are mostly available. For many operators in the informal economy, access to the export market is almost impossible due to issues of standards, packaging, and the huge capital requirements for such trade. In many countries in Africa, operators in the informal sector also face regulatory barriers such as complex registration processes, harassment by state officials both at the borders and within the countries, extortion, and multiple charges, as well as insecure property rights (Oloruntoba, 2018). Delays in the registration of business names affect the speed with which people who operate in this sector can have their businesses registered. As with many social services issues in Africa, there is an urban bias in registration points. In a country like Rwanda, where registration of business names has been made easier, the opportunity is not evenly distributed, as those in the rural areas still face significant hurdles to having their businesses registered. Lack of intentional support for operators in the rural areas remains a big challenge in unlocking the massive opportunities embedded in indigenous African economic activities.

Other major challenges in this sector include the lack of health insurance, the absence of pensions, and undue vulnerability to shocks. In the absence of health insurance, operators in this sector lack access to social protections that can result from sickness or disease, which can keep them away from work for a long time. This period away from work means loss of income and productivity. In their old age, lack of a pension makes them unduly exposed to chronic poverty. The only exception is those who have been able to train their children to a level where they can work and earn a decent income. In such instances, some of these children are charged with the responsibility of taking care of their aged parents.

Although communal living in the rural and peri-urban areas provides an opportunity for care, this is not a substitute for well-planned social protection programs and plans from which the people can benefit. To summarize, we have argued in this section that despite the huge potential that exists in African indigenous economic systems, framing this sector as informal has subjected it to undue pressure, exploitation, and abuse. The challenges facing the sector remain largely unaddressed because of the wrong framing. Attempts at formalizing this sector, which development institutions always advise African leaders to do, fall short of the real solutions. What is needed is not formalization per se but an intentional decolonial approach that requires listening and co-creation of strategies to understand what the people need to maximize the potential in this sector. Rather than trying to force the operators to formalize, the government should work with the people to address the aforementioned challenges in ways that ensure that the infrastructural gaps are filled, access to capital is guaranteed, social protections in the form of health insurance and pension are provided, and, where necessary, design training programs that can help them in the standardization of their products. Storage facilities should also be provided for agricultural products to ensure preservation. The above recommendations can only be useful if the grammar of informality is changed. To this we now turn in the next section.

Rethinking the Language of Economic Policy

The discussions around the dichotomization of economies into formal and informal are based on the Western language of economic policy. The motivation behind the ILO's categorization is in part informed by Keith Hart's study of the Ghanaian economy and the informal sector. The adoption of the term informal sector and, latterly, informal economy enables the understanding of African economies within their unique context. However, there is an undertone of Western economic language framing the discussion on the dichotomization of economies. Primarily, the idea that the path to economic development is linear and all economies need to follow that trajectory is a major trait of the modernization theory. This linear idea of development informs the argument that for African economies to develop, they need to shrink the informal economy. Particularly, the benchmarking of formalization as a prerequisite for the development of African economies echoes the narrative that dominant economic models are Western-centric and do not consider different contexts. These misconceptions follow a Western model of what work should be without considering how integral 'informal' economic activity is to the African continent. The ILO dichotomies depict that if something is not easily modelled after Western models, it is left out of the equation in mainstream economics. The home-based worker, the work-from-home personnel, and the woman trader mess up the model of productive labour established by mainstream economics, as they do not fit within the established norms. This forces us to acknowledge that dichotomies do not always work in certain contexts.

The idea of turning informality to formality requires us to critically engage the myopic lens through which the ILO views the informal economy in Africa.

Although there is an acknowledgement that most workers in the informal economy are voluntarily choosing to partake in it, the poverty-induced lens is the dominant understanding of why people choose to work in the informal economy. This narrative ignores the reality of the African context, wherein existence outside of formal economic structures has been integral to people's livelihood and economic sustenance. Trading, entrepreneurship, and self-employment are dominant in the African economic system, as far back as pre-colonial Africa (Akyeampong 2017), such that most Africans who engage in the informal economy voluntarily choose to partake in it. This reality is often lost in the Western economic language used to understand African economies.

Additionally, the insistence on formalization and inclusion of informal workers into the wage system ignores this reality. We cannot rethink the language of economic policy if we continue to adopt skewed economic models for understanding African economies while ignoring the context of the informal economy in most African countries. The idea that everyone wants to be part of the wage system and work formal jobs reduces the essence of the vast network of most African economies. This is particularly the case with IE, which places importance on development that must coexist with the protection of cultural values and traditions. It does not prioritize the maximization of profit but focuses on broader goals like self-determination, environmental sustainability, and community empowerment (Cheteni & Umejesi, 2024; Deka & Goswami, 2025; Padilla-Meléndez et al., 2022). Moreover, there is no agreed-upon understanding of what it means to formalize the informal economy, as there are still contestations on whether it means inclusion into the wage system, paying taxes, social protection, etc. What it tells us is that, while there is some agreement on the informality of most African economies, there is yet to be a standard model for formalizing them, which reverts to the discussion of modernization theorists that developing countries will have to follow the same path as developed countries. In this case, countries with a large informal economy will have to adhere to the same models that developed countries used before they can formalize their economies. This brings us back to the discussion around the Western-centrism of economic models and policy that continue to ignore contexts for sameness and uniformity. Western economic models have been proven to be fallible over the years; however, the perceived superiority of these models continues to be peddled globally amongst economists who argue that liberalism and Western economic models are the proven path to development, such that diverging models are debunked and not acknowledged because they do not fit within the established mainstream economic models.

Rethinking the language of economic policy, therefore, becomes a holistic endeavor that has to start with dismantling the dominance of mainstream

Western economic models that dictate what type of development is acceptable. This dismantling will have to engage the philosophy of economics and also an age-long debate on what development looks like and means. The engagement of these epistemological limitations will open the pathway for alternative economic models that recognize the reality of African economic models while expanding the frame of what it means for an economy to be formal or informal. In this regard, dominant economic models determine the economic policy and language that most international development agencies adopt; they decide which economy is developed, formal, or industrialized. Until the models underpinning this language are addressed, economic policy will continue to create bifurcation that regards African economies as mostly informal.

As previously noted, the framing of indigenous economic activities in Africa as informal is a direct result of epistemicide, which colonialism has fostered over the past two centuries. This framing has several implications. First, it delegitimizes indigenous economic activities, which have sustained Africans for generations. This delegitimization not only obscures reality but also reinforces coloniality of being, knowledge, and power (Ndlovu-Gathseni, 2018, 2013). It also robs the vast majority of people who are engaged in it of the sense of pride that is attributed to those who are in the formal sector. Second, by framing indigenous economic activities in Africa as informal, their massive contributions to the overall health of the economy in terms of employment generation and livelihood are underreported and undervalued. Yet as mentioned above, the so-called informal economy creates jobs for more than 80 percent of the population in several African countries. Beyond that, it also provides the resources for sustainable livelihood even for those who are fully employed in the formal sector. Resources from this sector have been used to train doctors, lawyers, engineers, and other professionals across the continent.

Thirdly, by using the grammar of informality to describe this sector, the support expected from the government is largely missing. Rather, harassment and extortion, especially of women who are engaged in cross-border trade, have been the order of the day (Nshimbi, Moyo, and Oloruntoba, 2018). By designating the African indigenous economic system as informal, the government has been unable to formulate appropriate policies that can help the operators in this sector to scale up. Using the broad brush of informality has led to lumping all manner of illicit and illegal activities together with legitimate economic activities that are indigenous to Africa, thus criminalizing a critical lifeblood of the economy. Changing the grammar of informality can offer new insights that can propel the government to provide support for the operators in this sector, without necessarily formalizing them. Support in the areas of seed funds, microcredits, and infrastructures such as regular and predictable energy supplies can help to unleash the massive potential for economic growth that is inherent in this sector. Additionally, by using appropriate language to describe this sector, cross-border traders can have better protection and support. This is because, despite the artificial borders that the colonialists created, which the postcolonial elites unfortunately retained, African traders, especially women, have continued to trade among themselves. But as we noted above, they have been subjected to all forms of harassment due to what scholars have referred to as the 'northern problem' (Ndlovu-Gatsheni and Mhlanga, 2013) or what we call coloniality of borders. Yet the massive trade networks that existed across Africa before the colonial disruptions provided a strong platform for cross-continent trade relationships and building a virile economic system (Ochonu, 2015).

Lastly, a change in the grammar of informality will affect how the international community views and design their intervention programs in Africa. For too long, they have adopted the modernization theory approach, which posits that African countries should follow the linear process of the states of economic growth as outlined by Rostow (1960). However, given the structural and historical differences between African and European societies, this approach has not worked to address the problem of poverty and inequality. As we noted above, the various interventions such as structural adjustment programs and other externally designed reform programs have rather led to a worsening of the economic situations on the continent. Rather than continuing to follow the worn-out strategies of designing economic blueprints that reinforce coloniality, a genuine process of co-creation and listening is imperative for reimaging how economic policies are designed in ways that incorporate indigenous economic systems in Africa. As the Yoruba saying goes, 'aki fa ori eni leyin eni', which means it is impossible to barb someone's head in his or her absence.

Working with the government, traditional business associations, community leaders, and operations of the informal economy to design economic policies that foreground the specific African indigenous economic system will help to address the problem of unemployment and poverty. Additionally, this can further help to address the challenges with data and statistics, which remain critical to appropriate planning. By using the language of informality to describe indigenous economic activities in Africa, the contributions of this sector to the overall economic performance are either underreported or not reported at all. This affects the extent to which African countries are classified by institutions such as the World Bank and the International Monetary Fund. This also affects the for a where African leaders can speak, as economic clubs like G7 or G20 are generally determined by the level of economic outputs measured in terms of the Gross Domestic Product.

Conclusion

This chapter examines the ILO categorization of economies as a precursor to rethinking the language of economic policy. It attempts a critique of dominant Western economic models that underpin most development projects across the world. We contend that the very idea that economies are categorized into formal and informal is a manifestation of dominant Western economic models. These models posit that economic activities should be organized in a certain way, and any economy that does not fit within the established frame is not acceptable. Discussions on informality and by extension, IE see it as a shortcoming, a consequence of failure that needs to be corrected.

Without understanding the specific context around which this informality and IE thrive and exist, efforts have been/are been made to shrink this informality into formality. The advocates, made up of international development institutions, consultants, and policy makers, believe that the informal economy poses a significant challenge to the development of most African states and therefore the transition from informal economy to formal is imperative. Western economists and the ILO are peddling this narrative and building programs to achieve this transition. What is clear in this narrative is that for a state to develop, it cannot have an informal economy. Thus, the informalization inherent in most African economies is seen as a setback that hinders growth and development. It also exposes the lack of understanding of the uniqueness of economic activities and means of sustenance that Africans engage in. It is based on this that this paper discusses the need to rethink the language of economic policy that bifurcates economies without contextual understanding.

We argue that the ILO categorization is based on Western economic models, and they do not acknowledge economic activities that Africans engage in. It calls for a rethinking; instead of the obsession with turning informality to formality, there needs to be a de-centering of mainstream Western economic models that dictate what type of development is acceptable. For instance, in IE, the development of the measurement of success is not necessarily profit. Insights from IE demonstrate that economic activity is also shaped by non-economic drivers, including social, cultural, and communal considerations. While the pursuit of individual profit remains present, it is balanced with broader community needs and collective objectives, which ultimately inform how success is defined and measured (Dana, 2015). Thus, this de-centering will enable the expansion of economic models, policy, and an informed language adopted to describe alternative models of economies. Finally, we must recognize that dichotomies are often created based on certain frames or benchmarks that may not necessarily be universally applicable. Rather than adopting dichotomies based on a provincialized framing, it is imperative to decolonize and de-provincialize the grammar of economic policy in Africa in ways that respect and foreground the originality and authenticity of Africa's Indigenous economic activities and its ongoing profound impacts on sustainable livelihood on the continent.

References

Adeniran, A. I. (2015). Informal trans-border trading and identity integration: Exploring the experiences of Nigerian women shuttle migrants in Côte d'Ivoire. *Buwa!, 2*(6), 23–28.

Ake, C. (1981). *A political economy of Africa.* Longman.

Akyeampong, E. (2017). *History of African trade.* African Export-Import Bank.

Alvarez, G., Tran, M., & Raina, R. (2011). *MDGs: Sub-Saharan Africa: Overcoming data gaps and ranking progress* (Discussion Paper No. 2011-01).

Asmal, Z., Bhorat, H., Lochmann, A., Martin, L., & Shah, K. (2014). *Supply-side economics of a good type: Supporting and expanding South Africa's informal economy* (CID Research Fellow and Graduate Student Working Paper No. 158).

Bauer, P. T. (1971). *Dissent on development: Studies and debates in development economics.* Cox and Wyman.

Chakrabarti, S., & Kundu, A. (2009). Formal-informal sectors' conflict: A structuralist framework for India. *Journal of Economic Development, 34*(2), 27–67.

Chen, M. (2012). *The informal economy: Definitions, theories and policies* (WIEGO Working Paper No. 1). Women in Informal Employment: Globalizing and Organizing.

Chen, M., Vanek, J., Lund, F., Heintz, J., Jhabvala, R., & Bonner, C. (2005). *Progress of the world's women 2005: Women, work and poverty.* UNIFEM.

Cheteni, P., & Umejesi, I. (2024). Indigenous knowledge in entrepreneurship and cultural tourism in rural areas. *Comparative Sociology, 23*(3), 340–359.

Deka, N., & Goswami, K. (2025). Indigenous entrepreneurship and agricultural development: Insights from the case of "Darjeeling tea" in Eastern Himalayas. *Climate and Development, 17*(10), 905–920.

Dana, L.-P. (2015). Indigenous entrepreneurship: An emerging field of research. *International Journal of Business and Globalisation, 14*(2), 158–169.

De Soto, H. (1989). *The other path: The economic answer to terrorism.* HarperCollins.

Easterly, W. (2009). How the Millennium Development Goals are unfair to Africa. *World Development, 37*(1), 26–35.

Feige, E. (2005). *Overseas holdings of U.S. currency and the underground economy* (Macroeconomics Working Paper No. 0501022). EconWPA.

Fosu, A. (2010). Africa's economic future: Learning from the past. *CESifo Forum, 11*(1), 62–71.

Fukuyama, F. (1992). *The end of history and the last man.* Free Press.

Gilman, N. (2003). *Mandarins of the future: Modernization theory in Cold War America.* Johns Hopkins University Press.

Hart, K. (1973). Informal income opportunities and urban employment in Ghana. *The Journal of Modern African Studies, 11*(1), 61–89.

Horn, Z. E. (2009). *No cushion to fall back on: The global economic crisis and informal workers.* WIEGO & Inclusive Cities.

International Labour Office. (1972). *Employment, incomes and equality: A strategy for increasing productive employment in Kenya*. International Labour Organization.
International Labour Office. (2002). *Decent work and the informal economy* (International Labour Conference, 90th Session Report). International Labour Organization.
International Labour Office. (2009). *The informal economy in Africa: Promoting transition to formality: Challenges and strategies*. International Labour Organization.
International Labour Office. (2010). *The informal economy and decent work: A policy resource guide*. International Labour Organization.
International Labour Office. (2013). *The informal economy and decent work: A policy resource guide: Supporting transitions to formality*. International Labour Organization.
International Labour Office. (2017). *Global employment trends for youth: Paths to a better working future*. International Labour Organization.
International Monetary Fund. (2017). The informal economy in Sub-Saharan Africa. In *Regional economic outlook: Sub-Saharan Africa*. International Monetary Fund.
Kapsos, S., & Bourmpoula, E. (2013). *Employment and economic class in the developing world* (ILO Research Paper No. 6). International Labour Organization.
Kiaga, A., & Leung, V. (2020). *The transition from the informal to the formal economy in Africa*. Background paper for GEPR Chapter 5.
Lewis, W. A. (1954). Economic development with unlimited supplies of labour. *The Manchester School, 22*(2), 139–191.
Lewis, W. A. (1967). Unemployment in developing countries. *The World Today, 23*(1), 13–22.
Mahon, R. (2017). Gendering development: The OECD's Development Assistance Committee, 1981–2000. In *The OECD and the international political economy since 1948* (pp. 335–356). Palgrave Macmillan.
Maloney, W. F. (2004). Informality revisited. *World Development, 32*(7), 1159–1178.
Martinussen, J. (2005). *Society, state, and market: A guide to competing theories of development*. Zed Books.
Masina, N. P. S., & Mgoduka, S. (2026). Challenges faced by informal sector entrepreneurs: A focus on the street vendors at the Mbombela Municipality. *Journal of Sustainable Development Law and Policy, 17*(2), 30–53.https://doi.org/10.4314/jsdlp.v17i2.2
Medina, L., Jonelis, A., & Cangul, M. (2017). *The informal economy in Sub-Saharan Africa: Size and determinants* (IMF Working Paper No. 17/156). International Monetary Fund.
Mhando, P. C., & Kiggundu, M. N. (2018). Introduction to special issue—Managing Africa's informal economy: Research, practice and advocacy. *Africa Journal of Management, 4*(3), 219–224.
Ndlovu-Gatsheni, S. J. (2018). *Decolonization, development and knowledge in Africa: Turning over a new leaf*. Routledge.
Ndlovu-Gatsheni, S. J. (2013). *Coloniality of power in postcolonial Africa: Myths of decolonization*. CODESRIA.
Ndlovu-Gatsheni, S. J., & Mhlanga, B. (2013). *Bondage of boundaries and identity politics in postcolonial Africa: The "Northern Problem" and ethno-futures*. Africa Institute of South Africa.
Noyoo, N. (2015). Introduction to the concepts of social development. In L. Calvelo, R. Lutz, & F. Ross (Eds.), *Development and social work*. Paulo Freire Verlag.
Noyoo, N. (2022). Critiquing Western development paradigms and theories in the age of the coronavirus (COVID-19): An African perspective. In M. Gonçalves, R. Gutwald, T. Kleibl, R. Lutz, N. Noyoo, & J. Twikirize (Eds.), *The coronavirus crisis and challenges to social development: Global perspectives*. Springer.

Nshimbi, C. C. (2018). Issues in African informality: What is the relevance for regional or continental integration? *Africa Insight, 48*(1), 41–61.

Nshimbi, C. C., Moyo, I., & Oloruntoba, S. O. (2018). Borders, informal cross-border economies and regional integration in Africa: An introduction. *Africa Insight, 48*(1), 3–11.

Obeng-Odoom, F. (2013). Africa's failed economic development trajectory: A critique. *African Review of Economics and Finance, 4*(2), 151–175.

Ochonu, M. E. (2015). The Wangara trading network in precolonial West Africa: An early example of Africans investing in Africa. In T. McNamee, M. Pearson, & W. Boer (Eds.), *Africans investing in Africa* (pp. 21–35). Palgrave Macmillan. https://doi.org/10.1057/9781137542809_2

Olaitan, Z. M. (2024, August 9). Indigenous knowledge-based entrepreneurship and the informal economy in Africa: Rethinking the grammar of economic policy. *African Indigenous Knowledge Research Network Blog.* https://africaindigenousresearch.com/indigenous-knowledge-based-entrepreneurship-and-the-informal-economy-in-africa-rethinking-the-grammar-of-economic-policy/

Olaitan, Z. M., & Oloruntoba, S. O. (2023). Beyond reaction: (Re)-imagining African agency in the decolonization of knowledge. In N. Andrews & N. Khalema (Eds.), *Decolonizing African studies pedagogies: Knowledge production, epistemic imperialism and Black agency.* Palgrave Macmillan.

Oloruntoba, S. O. (2016). Beyond the fetishism of gross domestic product: (Mis)governance and the challenges of poverty reduction in Nigeria. *Africa Insight, 46*(2), 1–17.

Oloruntoba, S. O. (2018). Regional integration and development in Africa: Rethinking borders and informality. *Africa Insight, 48*(1), 12–28.

Oloruntoba, S. O. (2025). Decolonizing knowledge on youth employment and entrepreneurship in Africa: Why the voices of Indigenous youth matter? In S. O. Oloruntoba, I. Moyo, & L. Zondo (Eds.), *Indigenous knowledge systems and youth entrepreneurship: Recentering the voices of marginalized communities in Africa* (pp. 15–27). Bloomsbury.

Ohnsorge, F., & Yu, S. (Eds.). (2022). *The long shadow of informality: Challenges and policies.* World Bank.

Padilla-Meléndez, A., Plaza-Angulo, J. J., Del-Águila-Obra, A. R., & Ciruela-Lorenzo, A. M. (2022). Indigenous entrepreneurship: Current issues and future lines. *Entrepreneurship & Regional Development, 34*(1–2), 6–31.

Poschke, M. (2019). Wage employment, unemployment and self-employment across countries. *SSRN Electronic Journal.* https://doi.org/10.2139/ssrn.3401135

Rostow, W. W. (1960). *The stages of economic growth: A non-communist manifesto.* Cambridge University Press.

Seers, D. (1979). *The meaning of development.* Routledge.

Standing, G. (1999). *Global labour flexibility: Seeking distributive justice.* St. Martin's Press.

The North Africa Post. (2026, April 18). *Africa's informal economy employs 83% of the workforce in 2024: UN data.* https://northafricapost.com/96624-africas-informal-economy-employs-83-of-workforce-in-2024-un-data.html

United Nations Economic Commission for Africa. (2014). *Report to the first joint session of the Committee of Directors General of the National Statistics Offices and the Statistical Commission for Africa on informal sector methodology in Africa.* https://archive.uneca.org/sites/default/files/uploaded-documents/Statistics/statcom2014/report_on_informal_sector_final_en.pdf

Chapter 4

INDIGENOUS KNOWLEDGE SYSTEMS AND NATURAL RESOURCES GOVERNANCE IN AFRICA

Introduction

Natural resource governance can be defined as the implementation of "hard and soft rules which shape and constrain" the exploitation and use of natural resources for the purpose of attaining and/or meeting the socioeconomic and sustainable development imperatives of a country, region or community that must benefit from such resources (Van Alstine, et al, 2014: 49). What is implicit in this definition is that natural resources governance must of necessity and design involve many actors including, the local, regional, national and those that are international (Van Alstine, et al, 2014). Another implicit assumption in the definition of natural resource governance is the idea that this must be done transparently to attain the intended consequences, such as socioeconomic and sustainable development. For instance, from about the 1990s, the resource curse paradox gained prominence to understand why countries that had huge amounts of natural resources were poorly developed (Sachs and Warner, 2001; Veit et al., 2011). Examining this paradox brought to the fore issues of transparency (see e.g., Van Alstine, et al, 2014; Benner et al., 2010), which was then linked to the resource curse or framed as bad governance leading to the intervention of international stakeholders to stop poor governance and corruption *inter alia*, especially in countries in the Global South (see e.g., Robinson et al., 2006; Van Alstine, 2014). A good example of this is the 1999 Global Witness Report, which chronicled the plunder and pillage of the Angolan natural resources during the civil war (Global Witness, 1999).

These transparency initiatives have also taken the form of the development of what are called good governance guidelines, like that of the World Bank, which sets guidelines on contracting, regulation of resource extraction operations, taxation and royalties, revenue generation and distribution, and attainment of sustainable development goals (SDGs) (see e.g., Alba,

2009). At face value, this comes across as good and beneficial, but, at a deeper and fundamental level, the local communities on whose land the resources are located and exploited do not benefit (see e.g., Buxton, 2012). It has been proposed that this can be mitigated by obligating companies to Corporate Social Responsibility (CSR) (Van Alstine, 2014). Nonetheless, there is also the recognition that CSR does not always solve the problem of transparency and beneficiation (see e.g., Cash, 2012). Another way to look at such initiatives is that they are simply effective ways of resource exploitation in countries in the Global South by multinational companies. It is tantamount to Eurocentric attempts aimed at reforming Eurocentrism and global capitalism. Such attempts can only succeed in protecting Eurocentrism and global capitalism by giving it a veil of legitimacy and acceptance, just like a Eurocentric critique of Eurocentrism is neither a decolonial posture nor a dismantlement of coloniality and Eurocentrism, but a strengthening of the same (see e.g., Alcoff, 2007, 2017; Maldonado-Torres, 2021). Differently stated, "what is required is a decolonizing of the way in which we interpret, and teach, Western philosophy and, indeed, every form of philosophical thought" (Alcoff, 2017: 399). Although this is not a contribution to philosophical thought, discussions in this chapter are located in this context by advancing that in Africa, one of the ways through which natural resources governance can be reformed so that it leads to inclusive and meaningful development is through the invocation of Indigenous methods and approaches. Applying Indigenous knowledge in the governance of natural resources is a manifestation of decoloniality, which is an epistemic resistance to the hegemony of Eurocentric perspectives. The chapter adopts a qualitative method based on a review of the literature and analysis of practices among Indigenous communities in Africa.

This is predicated on the fact that IKS are important in matters of self and community care as well as resource use and governance, for which reason they must be leveraged in ways that transcend tokenism for community and socioeconomic development. It is argued that the deployment of IKS in natural resource governance can create entrepreneurship opportunities that are grounded in community socioeconomic development. To take this debate forward, the following sections discuss the coloniality of natural resource governance in Africa and why and how this must be dismantled so that there are IKS-informed natural resource governance approaches to attain inclusive socioeconomic development in Africa. The final part of the chapter reflects on the imperative of reforming the current, exploitative, and inequitable methods of natural resource governance in Africa, which, far from dismantling coloniality, they entrench and consolidate it in hidden and sophisticated ways to the disadvantage of African people.

The Poverty of the Coloniality of Natural Resource Governance in Africa

In general, natural resources governance in Africa is state-led, and this takes many forms, such as the fortress conservation approach, which is a legacy of colonialism. This is because fortress conservation involves the creation of heavily guarded areas for the conservation of wildlife and ecosystems, and/or the creation of Protected Areas (PAs) to effectively fence off people so that they do not have access to natural resources. There is the assumption that when PAs are inaccessible to ordinary people and Indigenous communities, environmental degradation would not occur (see e.g., Brockington, 2002).

During colonial conquest and land grabbing by colonial settlers, many Indigenous communities in Africa, for example were displaced from and dispossessed of their lands based on the so called imperative to achieve resource governance and environmental management (see e.g., Brockington, 2002, 2004; Brockington and Schmidt-Soltau, 2004; Brockington and Igoe, 2006; Brockington and Wilkie 2015; Adams and Hutton 2007; Neumann 1998). To give these displacements a legal appearance, the colonizers applied the *terra nullius* principle. It is well recognized that *terra nullius* was a fabrication because, for instance, empty land did not exist in Africa. Fortress conservation strategies were and still are based on the forced removal of Africans from places where they have lived for generations (Brockington, 2002, 2004; Brockington and Schmidt-Soltau, 2004; Brockington and Igoe, 2006; Brockington and Wilkie, 2015; Adams and Hutton, 2007; Neumann, 1996, 1998; 2001, 2004a, b). The displacement of Black African people from their land has therefore meant that their IKS in terms of conserving the environment was discarded, and in its place were Eurocentric models and approaches to natural resources management and governance (see e.g. Ramutsindela, 2004; Moyo, 2023).

This is the context within which this has been correctly characterized as the imposition of wilderness on African people and the framing of Africa as a "lost Eden in need of protection and preservation" (Neumann, 1998:80). Indeed, Nelson (2004) has characterized this as environmental apartheid because it essentially replaced the epistemologies of Black African people regarding environmental governance and management with those that were completely foreign. The rationale for this is that Europe had to take the responsibility to protect Africa from Africans, and this entailed enforcing environmental apartheid and solidifying Eurocentric environmental conservation and governance epistemologies (see, e.g., 1998, 2001, 2002, 2004a, 2004b). There is nothing civilizational about this; if nothing else, it shows that colonialism and coloniality are "sick, decadent, and decivilizing" (Cesaire, 2000:31) and a declaration of war against humanity (Ndlovu-Gatsheni, 2018, 2020). Sadly, these colonial paradigms of natural resource governance and environmental management and conservation have persisted throughout the so-called post-colonial era (see, e.g., Adams and Mulligan, 2012). For example, the establishment of Protected Areas (PAs) is justified in many so-called post-colonial states by blaming Black Africans for environmental degradation (see, e.g., Erni, 2015; Mathews, 2003).

Furthermore, PAs should not be viewed as honest initiatives aimed at environmental governance and conservation; rather, they are also technologies that support colonial land ownership patterns and governance structures, as demonstrated by Ramutsindela (2004). The fact that many white landowners in Southern African countries like Botswana, Namibia, South Africa, and Zimbabwe used the concept of PAs to uphold their property rights just before these countries attained political independence supports this claim. Typically, agricultural farms were converted into PAs, which were then shielded from expropriation by the laws of the recently independent southern African governments. In other ways, agricultural land, which was quickly converted into PAs, could not be subject to redistribution, and ownership remained firmly in the hands of White people.

The effect of this is that the Indigenous people have continued to be marginalized and dispossessed even under the watch of so-called democratic governments (Ramutsindela, 2004). The point, therefore, is that Eurocentric approaches to natural resource governance and environmental management have been institutionalized based on the establishment of PAs (see, e.g., Brockington, 2002, 2004; Brockington and Schmidt-Soltau, 2004; Brockington and Igoe, 2006; Brockington and Wilkie, 2015; Adams and Hutton, 2007; Neumann,2004a 1998; , b). In other words, Protected Areas represent alternative means of maintaining colonial property control. It is even more disheartening that powerful, large international conservation NGOs are still pressuring the so-called post-colonial states to uproot Indigenous communities in the name of natural resource governance, environmental management, and conservation in the form of, among others, the creation of PAs (see, e.g., Cernea and Schmidt-Soltau, 2006). For this reason, there must be a change in the way that natural resources are governed if inclusive socioeconomic development can be attained in Africa.

Symbolic Inclusion of Indigenous Communities in State-Centric Governance of Natural Resources

In many African countries, there have been attempts at including Indigenous communities and ordinary people in the governance of natural resources in the form of, for example, co-management agreements in the governance of PAs. For example, after the democratic dispensation in South Africa, the government has recognized the importance of involving Indigenous communities in the governance of natural resources and the environment, leading to co-management agreements between the state and indigenous communities (see e.g., Musavengane and Leonard, 2019). Thus, co-management, also known as collaborative management, attempts to involve Indigenous communities in the governance, management, and conservation processes of the environment (Hill et al., 2012; Ward et al., 2018), for socioeconomic development and equitable distribution of natural resource access and utilization (Silva and Mosimane, 2014). It is for this reason that the co-management ought to lead to "pluralism, communication and negotiation, transactive decision-making, social learning," and "shared action/commitment" (Plummer and Fitzgibbon 2004:880). Put differently, the primary goal of co-management is to balance community needs with ecological sustainability (Kura et al., 2023). This approach contrasts with traditional fortress conservation strategies, which frequently marginalize and alienate Indigenous communities (Barletti and Rolando, 2024).

However, co-management has been correctly seen as symbolic because in reality there is neither power sharing nor fair distribution of resources and benefits (see, e.g., Calfucura, 2018; Musavengane, 2019; Kimengsi et al., 2019). In some cases, Indigenous communities do not have any say in how PAs are governed because the state continues to play a leading role, which effectively sidelines the Indigenous community members. In the case of Indigenous communities that were visited, leading to conversations and discussions, it becomes evident that such communities were not happy with the way that the so-called co-management agreement worked. The source of this unhappiness is that the co-management framework was in favor of the PA authorities in the first instance. The Indigenous communities needed to follow what was in the co-management plan, which they claimed they were not part of its development.

They therefore argued that the operation of the co-management framework was predicated on asymmetrical power relationships, which disempowered the local communities (also see Moyo, 2022). For the many indigenous communities in South Africa, this was problematic because, in the first instance, the PA was established on land on which they were displaced during colonial conquest, leading to the creation of the conservation area. Now that democracy was attained, there was the expectation that there would be meaningful involvement of the Indigenous communities in the governance of the PA and, by extension, the natural resources. This effectively meant that Eurocentric approaches to the governance of the natural resources and the

PA were the standard, and the views of Indigenous communities and their IKS were seen as secondary (also see Moyo, 2022).

Added to this are issues of corruption and lack of communication, which militate against the effectiveness and operation of co-management agreements (see, e.g., Calfucura, 2018; Musavengane, 2019). In other cases, co-management agreements have resulted in a dependency syndrome in that Indigenous communities completely rely on the state and other stakeholders for community development projects (Kimengsi et al., 2019). All this shows that the models of co-management have not fully addressed the issue of the governance of natural resources for the attainment of socioeconomic development in Africa. In all this, what remains unexplored is the role of IKS in the governance of natural resources for inclusive socioeconomic development. It needs reiteration that the colonial settlers displaced Indigenous communities and discarded IKS relevant to the governance of natural resources, and replaced them with Eurocentric approaches. Such Eurocentric approaches have failed and continue to lead to inclusive socioeconomic development in Africa. In cases where attempts have been made to reform or correct Eurocentric approaches like the co-management approach, things have not worked either. Perhaps, the time has come to relook at the potential of IKS in environmental governance and socioeconomic development in Africa. This is the logic that drives this chapter; hence, the next section examines case studies or examples where the deployment of IKS in the governance of the environment and natural resources has had positive outcomes.

IKS, Natural Resource Governance and Sustainability

In different parts of Africa, Indigenous communities have and continue to utilize IKS to sustainably govern and manage natural resources so that they (natural resources) continue to provide a livelihood base for future generations (see e.g, Obiero et al., 2022). This is different from extractive industries, which are mostly run by multinational corporations, most of which are based in North America and Western Europe. For instance, research by Ntiamoa-Baidu (1995) illustrated three ways in which IKS was used to conserve natural resources. First, IKS was used to protect specific ecosystems or habitats by constructing areas such as sacred groves, sacred forests, sacred sites, and rivers, for example. Second, IKS was utilized to conserve and manage specific animal or plant species using totems and taboos. In particular, the taboos were instrumental in controlling and monitoring access to and use of resources. Third, IKS were used to monitor or control the exploitation of natural resources, and these included the enforcement of closed seasons and closed days, which restricted the collection of natural resources through, for example, fishing and hunting. Such practices are still implemented by Indigenous communities like the Luo in Kenya, whose livelihoods depend on fishing in Lake Victoria (Opondo, 2016). Typically, people who engage in fishing have restrictions on when they fish and the methods that they use. Traditional leaders controlled access to the lake, and in this regard, closed seasons were observed and enforced. In addition, not all types and sizes of fish were open for fishing, because there were restrictions on specific species and sizes of fish that could be captured. The fishing itself involved traditional fishing methods like traps and spears, which had a limited impact on the fishery resource. Those who violated this were punished by the imposition of temporary fishing bans (Opondo, 2016). Likewise, in West Africa, catfish[1] fishing and farming in water sources that hosted catfish were regulated and monitored by Indigenous community leaders such as the Dogon people of Mali and the Akan people in Côte d'Ivoire, for example. The net effect of this is that the catfish was protected because of its sacred status. In some cases, the catfish was worshipped (Obiera et al., 2022). The effect of all these regulations and monitoring was that in cases of those fish that were open to fishing, overfishing was avoided. This is an example of sustainable fishing, which ensured that there was no depletion of fish, and this ensured that people could engage in fishing for their livelihoods and socioeconomic development over many generations (Opondo, 2016; Obiera et al., 2022).

In southern African countries like Zimbabwe, similar taboos were used to manage natural resources. A study by Reniko et al (2018) shows that among the Korekore ethnic group of Hurungwe, Zimbabwe, certain rare animal species such as pythons, pangolins, fish, and rhinoceroses are facing the threat of extinction. To address the unsustainable exploitation of these animals, the Korekore people have developed traditional conservation strategies rooted in taboos. Many birds and animals in the Hurungwe district are revered due to their perceived connection to spiritual forces. As such, it is a taboo to harm or kill wildlife deemed sacred. These taboos not only prohibit the exploitation of wildlife but also discourage cruelty towards animals. Through such IKS, the Korekore promote the idea that animals deserve moral consideration. Chiefs in the community enforce prohibitions against hunting certain species, including pangolins, baboons, monkeys, and pythons, as these animals are believed to embody spiritual beings. For example, the appearance of a solitary baboon is considered a significant omen within the Korekore community, symbolizing ancestral presence. Killing such a baboon is strictly forbidden, even if it is found in croplands. Similarly, zebras are protected because they are linked to spiritual beings, particularly since the local chief belongs to the Dube (zebra) totem. Killing a zebra would be seen as an affront to the chieftainship. Birds such as the *gondo* (eagle) are regarded as messengers of the ancestors, and harming them is culturally unacceptable (Reniko et al., 2018).

A closer examination of these beliefs reveals that their purpose is environmental governance and conservation. Essentially, the Korekore people's reverence for these animals forms a traditional and spiritually grounded system of wildlife protection and environmental stewardship (Reniko et al., 2018). In addition to safeguarding water sources and wildlife, some taboos among the Korekore ethnic group were aimed at protecting natural vegetation. Certain plants and landscapes are revered as habitations of spiritual forces, and as such, it is a taboo to cut them down or exploit them indiscriminately. Sacred sites, such as specific forests and mountains, are particularly protected. These areas are considered spiritually significant, and members of the Korekore community are discouraged from entering them or disturbing their natural state. Cutting down trees or desecrating these consecrated places is strictly forbidden, as doing so is seen as a violation of cultural and spiritual values. These taboos reflect a belief in the intrinsic value of vegetation, beyond its usefulness to humans (Reniko et al., 2018).

Similar practices have been recorded in countries like Uganda. For instance, the Lugbara community in Yumbe district, northwestern Uganda, also used taboos and totems to protect their natural resources. In this region, IKS was used to conserve and manage different types and varieties of vegetation for community and environmental wellbeing (Alule et al., 2023). Examples of these include big trees, trees that provided food, medicine, honey, and those that grew in sacred areas. Regarding big trees, they were preserved based on the belief that they were inhabited by powerful spirits and cutting them down could provoke these spirits and bring misfortune. They served as natural landmarks, used as directional compasses and as boundary markers between tribes, clans, sub-clans, and households. These trees also functioned as central gathering spots for cultural and communal events. Additionally, it was believed that large trees stored significant amounts of water, which helped regulate rainfall and ensured a steady flow of surface water in streams, wells, and springs. The second category of protected trees included those that provided food, such as fruits, tubers, leaves, bark, and seeds. Examples of these are the shea nut tree (*komoro*), tamarind (*itii*), *lugba*, and *lo'boni*. These trees were not to be cut down, as they were considered vital sources of sustenance, especially during times of famine. The third category comprised trees used for herbal medicine, such as mahogany (*marigo*), *izekize*, *galabalogoro*, and *gbulikigbuli*. These were to be harvested with care since their loss was believed to threaten the health and well-being of the entire community. The fourth category included ancient trees that housed beehives. These trees were preserved not only for their honey, which is a valuable food source, but also for the benefit of future generations. Finally, concerning the trees located within sacred grounds, these areas were accessible only to certain elders or specific clan members and were often rich in biodiversity. This richness in plant and animal life was thought to stem from the deep spiritual reverence and fear associated with disturbing these sacred spaces (Alule et al., 2023). Furthermore, Indigenous communities demonstrated an ethical responsibility towards the environment and its natural resources. In a study of the Me'en community in Ethiopia, Berhane (2024) indicates that, individually and collectively, Indigenous communities had a responsibility towards conserving and protecting natural resources not only for the current and future generations, but also because they believed that it was an ethical responsibility to do so. Stated differently, in the Me'en Indigenous community in Ethiopia, there exists a strong sense of virtue ethics at both the individual and collective levels. People tend to view the unauthorized use, unnecessary damage, or pollution of natural resources as not only inappropriate, but also dishonorable to the character of the person responsible and to the dignity of the environment and the whole society. There is also a deeply held belief regarding the fact that it is a privilege to be a custodian of the earth and its resources, and this places the ethical and moral duty to carefully manage and use them (Berhane, 2024).

However, the introduction of Western approaches to natural resource management has put pressure on IKS and, in some cases, has led to environmental degradation, which endangers the natural resources and livelihoods of Indigenous communities (Obiera et al., 2022). This is the context within which it has been correctly argued that IKS can proffer a valid, context-specific, and useful knowledge to enhance our understanding of how best natural resources can be governed to meet the imperative of sustainable conservation and socioeconomic development (Tengö et al., 2014; Alule et al., 2023). Hence, we argue that there is a need to rethink the governance of natural resources in Africa in ways that make IKS an integral component. This will maintain the stability of the environment but also lead to development that is inclusive because the people will be an integral part of the governance of resources to meet their needs (see e.g., Obiera et al., 2022). It is the Indigenous communities who know and understand their environment and needs, and hence the necessity for their inclusion in any natural resource governance initiatives. This is because IKS demands group action and operates based on mutual regulation that local communities have evolved over many centuries (see e.g., Reniko et al., 2018). This has been illustrated in the preceding illustration of several examples in east, southern and west Africa in which Indigenous communities practiced IKS in respect of natural resource governance. For instance, in the case of the Lugbara community in northwestern Uganda, group and community action were central in the governance and management of the environment. For example, the trees that provided food were cared for by the community, and the corollary is that the benefits were also for the community, and any socioeconomic development integral to this was inclusive.

Similarly, those trees that provided medicine were taken care of by the community, and the health benefits arising out of this were enjoyed by the community. This suggestion is that by using IKS, the governance of natural resources is geared towards the benefits of the entire community and not just individuals. It is our contention that this IKS principle shows that natural resources in Africa can be governed in ways that benefit all people who have a claim or share in natural resources in a specific territory. This principle also suggests that the current governance of natural resources in Africa, in ways that exclude local communities, must be reformed because it is possible to infuse the current methods of natural resource governance with those of Indigenous communities for both environmental sustainability and inclusive socioeconomic development. Any so-called environmental governance or management that excludes local communities and/or Indigenous communities cannot lead to authentic and inclusive socioeconomic development.

Regarding the Rigor of Indigenous Knowledge Systems in Natural Resource Governance

Considering the preceding discussion, it is necessary at this point to address concerns that have been raised about the efficacy of IKS in natural resources and environmental governance. For example, Smith and Wishnie (2000) have argued that IKS cannot be effectively used to govern natural resources. For these reasons, they have maintained that only trained ecologists and environmental managers who are deemed to know better can manage and conserve natural resources (Smith and Wishnie, 2000). Linked to this is the view that IKS is limited in its application because it is not scientific enough to detect environmental problems that are not visible and observable (see e.g. Beinart and Brown 2013; Murata et al. 2019). These concerns must be dismissed because they are fatally flawed. First, any suggestion on the inadequacy of IKS in environmental governance suggests that it can only be effective if it is integrated into Western science, and this is problematic because such integration is based on unequal power relationships. Western science is considered superior knowledge, and IKS is seen as inferior (Nadasdy 1999). Integration based on unequal power relations tends to extend the reach of Western scientific resource management into local communities in ways that reinforce the control of the state and its functionaries. Instead of genuinely merging two distinct knowledge systems, this approach often consolidates power in the hands of scientists and resource managers. Far from empowering local communities, it shifts authority and decision-making away from them, further marginalizing their knowledge and agency in favor of the elites and state (Nadasdy 1999). Furthermore, the idea of integrating IKS into Western science based on asymmetrical power relations explains why many so-called scientists, particularly resource managers, have been hesitant to recognize the existence of IKS. Instead, by framing natural resource governance strictly in terms of formal Western scientific governance and management systems, this perspective ultimately serves to legitimize and extend the reach of scientific resource management institutions into local communities (Nadasdy 1999). This leads to "knowledge making practices [that] marginalise or discount specific people and places, especially by privileging what might be generalized as 'Western' (often Euro-white) knowledge over 'southern' (often racialized) ways of seeing and knowing the world" (De Leeuw and Hunt 2018:3). In this chapter, we advocate an end to this by pushing for the recognition of IKS in environmental governance.

Second, discussions in the preceding have shown that Indigenous communities, who are the custodians of IKS, do not need training in the Western-centric sense. They have been trained enough through living in and interacting with nature over many generations. They do not need to be called ecologists or environmental managers to be qualified to govern and manage natural resources. These communities have deeply rooted social and cultural values, supported by structured systems of social control and cohesion. These systems are grounded in their IKS; both of the universe at large and of their specific local environments. Their interactions with the broader ecological system reflect well-defined, normative behaviors, shaped by a worldview that emphasizes harmony, adaptation, and respect for the natural world (Berhane, 2024). They have successfully done this in the past, and this does not need to be validated by Eurocentric science. For this reason, they do not need to have gone through the laboratory and put on a white laboratory coat to prove their knowledge. In any event, why must Eurocentric training be considered as the training and not any other? More than that, to suggest that Indigenous communities are not trained enough should be seen for what it is: Eurocentric heteronormativity, which regards Western science as the only knowledge. This must be resisted and rejected, for Europe and its knowledge is part of the world and not the world (see Ndlovu-Gatsheni, 2020; 2015; 2013; Maldonado-Torres, 2007; Wynter, 2003).

This is why in this chapter we foreground the necessity of rethinking the whole idea of natural resource governance, which is underpinned by Eurocentric logic and epistemology. We argue for natural resource governance to have IKS as an integral point of reference, which would lead to the exploitation and use of resources in ways that lead to inclusive socioeconomic development. In making this proposition, there is no intention to romanticize IKS, but to suggest that it is one of the epistemologies of people in Africa and the Global South, which must be seriously remembered and written back into matters of natural resource governance and socioeconomic development. In any event, IKS and Western epistemology on natural resources governance can complement, but this should be based on symmetrical rather than the past and present asymmetrical relationship, especially given the need for the coproduction or cocreation of knowledge for the betterment of environmental governance and management (see e.g. Moyo, 2022; Yanou et al, 2023). This point is taken forward in the next section, which demonstrates examples of the use of IKS in natural resource governance in ways that lead to entrepreneurial ventures leading to inclusive socioeconomic development.

The Case of IKS, Entrepreneurship, and Economic Development in Africa

The application of IKS on governance of natural resources in different parts of the world suggests that there is potential for entrepreneurship and economic development. For example, a study of Indigenous natural resource governance in the Southwest Amazon (Apurinã territories) showed that not only were they effective in preserving forest cover, but also had socioeconomic development impacts as compared to those territories that were under state control and governance (Virtanen et al., 2025). Despite the fact that these territories were under the immense pressure of deforestation, they were effective in responding to such pressure and successfully repelling deforestation due to the complex Indigenous methods. Briefly, each Apurinã territory was constituted of villages or communities led by a traditional leader. There were also traditional elders (kiiumanetxi), shamans (mỹyty), and several classes of knowledge-holders, like singers and storytellers, who were the custodians of Apurinã history, culture, and traditions.

More recently, these have been complemented by Apurinã teachers, environmental advocates, health and similar practitioners (Virtanen et al., 2025; Veber and Virtanen, 2017). In addition to the Indigenous political movement, the Apurinã regions are connected to one another by long-standing alliances and endogamous marriages amongst various clans. The only non-Indigenous people who reside in the Apurinã areas are those who have married an Apurinã, usually a woman. By using and caring for their lands across generations, Apurinã families can use resources in the area collectively, such as gathering Brazil nuts. Therefore, decisions are taken at different levels, including village, territorial, and between-territorial, as well as with other Indigenous countries. In addition to sharing knowledge and information in more informal networks, the communities hold village and territory meetings to deal with governance issues (Virtanen et al., 2025).

The economic activities undertaken by the Apurinã differ from the nearby extractive industries because they employ intricate, long-term cyclical ecological techniques that strongly support a rich biodiversity, like rotating agriculture and keeping the use of natural resources at a level that is sufficiently regenerative (Virtanen et al., 2025). Forest-covered areas are used for Apurinã economic activities like hunting, collecting, and fishing, as well as traditional family-based businesses like harvesting copaiba, Brazil nuts, and other natural oils and making crafts. The monitoring and protection of territories from illegal activities is a collective community effort, the result of which is that every member of the community takes part in the natural governance structures in this way. The Apurinã have maintained their traditional economic activities, such as the traditional small-scale swidden agriculture, the selling of Brazilian nuts and palm fruits, as well as art and craftwork. The products from these economic activities are sold in many markets in the territories. All these economic activities are forest-based, and yet they assist in providing incomes and other socioeconomic development needs (Virtanen et al., 2025). More than that, a common view of the forest as a system that sustains life is the foundation of the reciprocal subsistence economies of the Apurinã, which are founded on relational values.

Furthermore, in the Apurinã areas, which exhibit high levels of deforestation mitigation and/or avoidance, governance and socio-bioeconomics are practiced. Apurinã's ecological traditions and land management systems are based on multi-temporal and multi-relational perspectives (Virtanen et al., 2025). The example of Apurinã illustrates a case where IKS has been used to govern the community and natural resources in ways that do not destroy and degrade the environment but regenerate it so that people can undertake forest-based economic activities from which they sell products in the market. These small enterprises can be classified as Indigenous entrepreneurship. To this extent, a link or connection between IKS, natural resource governance, and entrepreneurship can be seen. The point being made here is that the example of Apurinã shows that IKS can be effectively used in the governance of natural resources in ways that lead to socioeconomic development. While the Apurinã did live according to the Brazilian laws, at a local level, the governance of natural resources occurred according to IKS and Indigenous methods, leading to successful conservation and protection of the environment. This is striking because the adjacent territories, which were subjected to extractive industries, had witnessed debilitating deforestation, and yet the Apurinã had avoided deforestation in their territories based on Indigenous methods of natural resources governance (Virtanen et al., 2025). It is these territories that had avoided deforestation from which people extracted materials and made goods for sale in the markets, and which contributed to socioeconomic development across the whole communities. This shows that Eurocentric approaches to resource extraction, despite their sophisticated machinery, continue to fail to save the environment and achieve inclusive socioeconomic development.

All this shows that in the contemporary conjecture, there is a space for IKS and indigenous methods in natural resource governance for socioeconomic development. If this were not the case, how else could the faster deforestation in areas that were subjected to extractive industries and in which IKS were discarded be explained vis-à-vis those territories that avoided deforestation because of the use of IKS and Indigenous methods in natural resource and environmental governance? In this sense, the example of Apurinã shows how IKS in the governance of natural resources does lead to socioeconomic development that is inclusive to the extent that the whole community benefits. Even more importantly, is the fact that such economic activities do not occur at the expense of but conserve the environment, so that the same economic activities are sustainable. This deserves commentary in the following section.

The Imperative of Decolonizing Natural Resource Management: Toward Socioeconomic Development in Africa

In this chapter, the attempt was made to establish the link between natural resource governance, entrepreneurship, and socioeconomic development. By bringing to bear case studies and examples from different parts of Africa and the Global South, it is evident that IKS must be leveraged in the governance of natural resources. This is because one of the central principles that underpin the governance of natural resources using IKS is collective action, decision-making, accountability, and benefit. The point being made is that IKS are important in matters of self and community care as well as natural resource use and governance, for which cause these practices can be leveraged for community and socioeconomic development in Africa. For example, the governance of natural resources using IKS in many parts of Africa shows that IKS instills and inculcates values of community and less individual development and success. This can be further illustrated by visits to and conversations with Indigenous communities in South Africa, who indicated that if the organizing logic in the governance of natural resources were IKS, it would be possible to initiate entrepreneurship projects that lead to inclusive socioeconomic development based on the community and less individual focus of such entrepreneurship. The Indigenous communities argued that their exclusion from the governance of natural resources in PAs was one of the reasons why there was widespread poverty despite the existence of resources whose benefit they did not enjoy. They argued that the neoliberal model around the governance of natural resources in PAs meant that the people who should have benefited or should be benefiting were marginalized.

In other words, the Indigenous communities wanted to be involved in the governance structure of PAs so that they could have access to natural resources out of which they could form viable businesses (also see Moyo, 2022). Indeed, this chapter has discussed several examples of Indigenous communities in Africa in which the governance of natural resources was done sustainably using taboos and totems, for example. This was effective in terms of regulating and monitoring which resources were used, how, and by whom. This immediately dismisses any claim that Indigenous communities or their IKS cannot be used to govern and manage the environment. For example, some Indigenous community members involved in the making of grass mats knew better ways of harvesting the grass, and this did not need any Western science or experiments. Thus, denying them access to the PAs did not lead to any better conservation or the upliftment of the community. Therefore, if they were involved in the governance of these resources, they would still harvest the grass out of which they made the mats in ways that protected the environment. They would then form businesses based on grass mat making, out of which individuals and communities would lay claim to decent livelihoods. This is not a utopian proposition because, as shown by the case of the Apurinã discussed in this chapter, using IKS to govern natural resources led to Indigenous entrepreneurship, which was based on forest resources, and this contributed to incomes and other socioeconomic development imperatives (Virtanen et al., 2025).

What is even more noteworthy is that Indigenous entrepreneurship used simple resources from the environment in ways that were sustainable. Indigenous entrepreneurship led to the creation of value from community-based resources such as people, land, skills, and culture and reinvesting that value in ways that support the socioeconomic development of Indigenous communities (e.g., Colbourne et al., 2023; Colbourne, 2017; Peredo and Anderson, 2006). Consequently, it is important that IKS is leveraged and unleashed towards the formation of Indigenous entrepreneurship because it (Indigenous entrepreneurship) is the cornerstone of the contemporary indigenous economy. Indigenous entrepreneurship is both the answer and the path forward for the development and cultural survival of African communities. It is important to remember that Indigenous communities in

Africa and Latin America have a long history of business ownership and operations, including, among others, engaging in barter trading (Cheteni and Umejesi, 2024; Munawar et al., 2020). More than that, Indigenous entrepreneurship is correctly seen as a good way through which Indigenous communities can reconstruct themselves after many centuries of colonization, which was accompanied by the erosion and/or degradation of the traditional epistemological, social, ecological, cultural, and other resources and identities (Murphy et al., 2020). For this reason, the chapter advocates for the reform of the governance mechanisms of natural resources among the Indigenous communities in Africa. This chapter contributes to the discourse by highlighting contemporary governance gaps in youth entrepreneurship policy responses and advocating for inclusive employment opportunities at the community level.

Note

1 The catfish was sacred

References

Adams, W. M., & Hutton, J. (2007). People, parks and poverty: Political ecology and biodiversity conservation. *Conservation and Society, 5*(2), 147–183.

Adams, W., & Mulligan, M. (2012). Introduction. In W. Adams & M. Mulligan (Eds.), *Decolonizing nature: Strategies for conservation in a post-colonial era* (pp. 1–15). Earthscan.

Alba, E. M. (2009). *Extractive industries value chain* (Extractive Industries for Development Series No. 125). World Bank. https://documents1.worldbank.org/curated/en/282401468339611763/pdf/484240NWP0Box31ei1for1development13.pdf

Alcoff, L. M. (2007). Mignolo's epistemology of coloniality. *CR: The New Centennial Review, 7*(3), 79–101.

Alcoff, L. M. (2017). Philosophy and philosophical practice: Eurocentrism as an epistemology of ignorance. In I. J. Kidd, J. Medina, & G. Pohlhaus Jr. (Eds.), *The Routledge handbook of epistemic injustice* (pp. 397–408). Routledge.

Alule, J. R., Nuwategeka, E., Oriangi, G., & Lajul, W. (2023). Relevance of indigenous knowledge in sustainable management of forest resources in the 21st century Uganda. *East African Journal of Environment and Natural Resources, 6*(1), 277–296.

Barletti, J. P. S., & Rolando, G. (2024). Between co-management and responsibilisation: Comparative perspectives from two reservas comunales in the Peruvian Amazon. *Bulletin of Latin American Research, 43*(2), 104–119.

Beinart, W., & Brown, K. (2013). *African local knowledge and livestock health: Diseases and treatment in South Africa.* Wits University Press.

Benner, T., de Oliveira, R. S., & Kalinke, F. (2010). The good/bad nexus in global energy governance. In *Global energy governance: The new rules of the game* (p. 372).

Berhane, A. T. (2024). Indigenous knowledge of Me'en community toward natural resource conservation: A socio-philosophical analysis. *AlterNative: An International Journal of Indigenous Peoples, 20*(1), 55–66.

Brockington, D. (2002). *Fortress conservation: The preservation of the Mkomazi Game Reserve, Tanzania.* James Currey.
Brockington, D. (2004). Community conservation, inequality and injustice: Myths of power in protected area management. *Conservation and Society,* 2, 411–432.
Brockington, D., & Igoe, J. (2006). Eviction for conservation: A global overview. *Conservation and Society, 4,* 424–470.
Brockington, D., & Schmidt-Soltau, K. (2004). The social and environmental impacts of wilderness and development. *Oryx, 38,* 140–142.
Brockington, D., & Wilkie, D. (2015). Protected areas and poverty. *Philosophical Transactions of the Royal Society B: Biological Sciences, 370*(1681), 1–6.
Buxton, A. (2012). *MMSD+10: Reflecting on a decade of mining and sustainable development.* International Institute for Environment and Development. https://www.iied.org/sites/default/files/pdfs/migrate/16041IIED.pdf
Calfucura, E. (2018). Governance, land and distribution: A discussion on the political economy of community-based conservation. *Ecological Economics, 145,* 18–26.
Cash, A. C. (2012). Corporate social responsibility and petroleum development in Sub-Saharan Africa: The case of Chad. *Resources Policy, 37*(2), 144–151.
Cernea, M. M., & Schmidt-Soltau, K. (2006). Poverty risks and national parks: Policy issues in conservation and resettlement. *World Development, 34,* 1808–1830.
Césaire, A. (2000). *Discourse on colonialism.* Monthly Review Press.
Cheteni, P., & Umejesi, I. (2024). Indigenous knowledge in entrepreneurship and cultural tourism in the rural areas. *Comparative Sociology, 23*(3), 340–359.
Colbourne, R. (2017). Indigenous entrepreneurship and hybrid ventures. In A. C. Corbett & J. A. Katz (Eds.), *Hybrid ventures* (pp. 93–149). Emerald Publishing.
Colbourne, R., Peredo, A. M., & Henriques, I. (2023). Indigenous entrepreneurship? Setting the record straight. *Business History, 66*(2), 455–477.
De Leeuw, S., & Hunt, S. (2018). Unsettling decolonizing geographies. *Geography Compass, 12*(7), 1–14.
Erni, C. (2015). *Shifting cultivation, livelihood and food security: New and old challenges for Indigenous peoples in Asia.* Food and Agriculture Organization of the United Nations.
Global Witness. (1999). *A crude awakening: How Angolan state corruption and the lack of oil company and banking transparency has contributed to Angola's humanitarian and development catastrophe.* Global Witness.
Hill, R., Grant, C., George, M., Robinson, C. J., Jackson, S., & Abel, N. (2012). A typology of Indigenous engagement in Australian environmental management: Implications for knowledge integration and social-ecological system sustainability. *Ecology and Society, 17*(1), 23–41.
Kimengsi, J. N., Aung, P. S., Pretzsch, J., Haller, T., & Auch, E. (2019). Constitutionality and the co-management of protected areas: Reflections from Cameroon and Myanmar. *International Journal of the Commons, 13*(2), 1003–1020.
Kura, Y., Mam, K., Chea, S., Eam, D., Almack, K., & Ishihara, H. (2023). Conservation for sustaining livelihoods: Adaptive co-management of fish no-take zones in the Mekong River. *Fisheries Research, 265,* 1–11.
Maldonado-Torres, N. (2007). On coloniality of being: Contributions to the development of a concept. *Cultural Studies, 21*(2–3), 240–270.
Mathews, A. S. (2003). Suppressing fire and memory: Environmental degradation and political restoration in the Sierra Juarez of Oaxaca, 1887–2001. *Environmental History, 8,* 77–108.

Moyo, I. (2022). Beyond a tokenistic inclusion of indigenous knowledge systems in protected area governance and management in Okhahlamba-Drakensberg. *African Geographical Review, 42*(2), 141–156.

Murata, C., Mantel, S., De Wet, C., & Palmer, A. R. (2019). Lay knowledge of ecosystem services in rural Eastern Cape Province, South Africa: Implications for intervention project planning. *Water Economics and Policy, 5*(2), 1–29.

Murphy, M., Danis, W. M., & Mack, J. (2020). From principles to action: Community-based entrepreneurship in the Toquaht Nation. *Journal of Business Venturing, 35*(6).

Musavengane, R. (2019). Using the systemic-resilience thinking approach to enhance participatory collaborative management of natural resources in tribal communities: Toward inclusive land reform-led outdoor tourism. *Journal of Outdoor Recreation and Tourism,* 25, 45–56.

Musavengane, R., & Leonard, L. (2019). When race and social equity matter in nature conservation in post-apartheid South Africa. *Conservation and Society, 17*(2), 135–146.

Nadasdy, P. (1999). The politics of TEK: Power and the "integration" of knowledge. *Arctic Anthropology, 36*(1–2), 1–18.

Ndlovu-Gatsheni, S. J. (2013). Why decoloniality in the 21st century? *The Thinker, 48,* 10–15.

Ndlovu-Gatsheni, S. J. (2015). Decoloniality in Africa: A continuing search for a new world order. *ARAS, 36*(2), 22–50.

Ndlovu-Gatsheni, S. J. (2018). *Epistemic freedom in Africa: Deprovincialization and decolonization.* Routledge.

Ndlovu-Gatsheni, S. J. (2020). *Decolonization, development and knowledge in Africa: Turning over a new leaf.* Routledge.

Nelson, H. N. (2003). Environmental colonialism: "Saving" Africa from Africans. *The Independent Review, 8*(1), 65–86.

Neumann, R. P. (1996). Dukes, earls and ersatz Edens: Aristocratic nature preservationists in colonial Africa. *Environment and Planning D: Society and Space, 14,* 79–98.

Neumann, R. P. (1998). *Imposing wilderness: Struggles over livelihood and nature preservation in Africa.* University of California Press.

Neumann, R. P. (2001). Africa's "last wilderness": Reordering space for political and economic control in colonial Tanzania. *Africa, 71,* 641–665.

Neumann, R. P. (2004a). Nature-state-territory: Towards a critical theorization of conservation enclosures. In R. Peet & M. Watts (Eds.), *Liberation ecologies: Environment, development, social movements* (pp. 195–217). Routledge.

Neumann, R. P. (2004b). *Making political ecology.* Hodder Arnold.

Ntiamoa-Baidu, Y. (1995). *Indigenous vs. introduced biodiversity conservation strategies: The case of protected area systems in Ghana.* WWF Biodiversity Support Program.

Obiero, K. O., N'Guessan, S. K., Migeni, A. Z., & Achieng, A. O. (2022). Bridging Indigenous and non-Indigenous knowledge systems and practices for sustainable management of aquatic resources from East to West Africa. *Journal of Great Lakes Research, 49*(1), S128–S137.

Opondo, P. (2016). Fisheries as heritage: Indigenous methods of fishing and conservation among the Luo fishers of Lake Victoria, Kenya. In A.-M. Deisser & M. Njuguna (Eds.), *Conservation of natural and cultural heritage in Kenya: A cross-disciplinary approach* (pp. 200–211). UCL Press.

Peredo, A. M., & Anderson, R. B. (2006). Indigenous entrepreneurship research: Themes and variations. In C. S. Galbraith (Ed.), *Developmental entrepreneurship: Adversity, risk, and isolation* (pp. 253–273). Emerald Publishing.

Plummer, R., & FitzGibbon, J. (2004). Co-management of natural resources: A proposed framework. *Environmental Management, 33*, 876–885.

Ramutsindela, M. (2004). *Parks and people in postcolonial societies: Experiences in Southern Africa*. Springer.

Reniko, G., Mogomotsi, P. K., & Mogomotsi, G. E. J. (2018). Integration of indigenous knowledge systems in natural resources management in Hurungwe District, Zimbabwe. *International Journal of African Renaissance Studies, 13*(1), 96–112.

Robinson, J. A., Torvik, R., & Verdier, T. (2006). Political foundations of the resource curse. *Journal of Development Economics, 79*(2), 447–468.

Sachs, J. D., & Warner, A. M. (2001). The curse of natural resources. *European Economic Review, 45*(4–6), 827–838.

Silva, J. A., & Mosimane, A. (2014). How could I live here and not be a member? Economic versus social drivers of participation in Namibian conservation programs. *Human Ecology, 42*(2), 183–197.

Smith, E. A., & Wishnie, M. (2000). Conservation and subsistence in small-scale societies. *Annual Review of Anthropology, 29*, 493–524.

Tengö, M., Brondizio, E. S., Elmqvist, T., Malmer, P., & Spierenburg, M. (2014). Connecting diverse knowledge systems for enhanced ecosystem governance: The multiple evidence base approach. *AMBIO, 43*, 579–591.

Van Alstine, J., Manyindo, J., Smith, L., Dixon, J., & Ruhanga, I. A. (2014). Resource governance dynamics: The challenge of "new oil" in Uganda. *Resources Policy, 40*, 48–58.

Veit, P. G., Excell, C., & Zomer, A. (2011). *Avoiding the resource curse: Spotlight on oil in Uganda* (WRI Working Paper). World Resources Institute.

Ward, C., Stringer, L. C., & Holmes, G. (2018). Protected area co-management and perceived livelihood impacts. *Journal of Environmental Management, 228*, 1–12.

Wynter, S. (2003). Unsettling the coloniality of being/power/truth/freedom: Toward the human, after man, its overrepresentation—An argument. *CR: The New Centennial Review, 3*(3), 257–337.

Yanou, M. P., Ros-Tonen, M. A. F., Reed, J., Moombe, K., & Sunderland, T. (2023). Integrating local and scientific knowledge: The need for decolonising knowledge for conservation and natural resource management. *Heliyon, 9*(11), 1–18.

Chapter 5

INDIGENOUS ENTREPRENEURSHIP AND FINANCING: A CASE STUDY OF ESUSU

Introduction

Entrepreneurship has been acknowledged to be integral to Africa's socioeconomic development (Ochonu, 2016; Olaitan, 2024). It provides for wealth creation while also addressing poverty and unemployment rates on the continent. Additionally, engagement in entrepreneurship is common amongst Africans, as many see it as a legitimate economic activity outside of conventional or 'formal' economic activities. Understanding how broader entrepreneurship contributes to socioeconomic development paves the way for a niche understanding of Indigenous entrepreneurship and its contextual nuances. With the increased interest in entrepreneurship as a contributor to Africa's development, there are two issues to note. First, there is a gendered nuance to the discourse of entrepreneurship on the continent, i.e., women form the majority in the entrepreneurial space, demonstrating the gendered nature of entrepreneurship in Africa and the challenges women entrepreneurs face (Morsy and El-Shal 2019). Second, most entrepreneurial activities that Africans engage in are related to Indigenous entrepreneurship because these activities are rooted in cultures and traditions. For instance, the woman trader who sells clothes in the market exemplifies an Indigenous entrepreneur due to the historical attachment of her trade to her community. Drawing from this example, women make up a majority of Indigenous entrepreneurs across the continent as they find not only economic empowerment but also social belonging through their work (Simba et al. 2023). We can recognize that women make up a majority of entrepreneurs in Africa, while not ignoring the many challenges that they face in the course of their economic activity. Entrepreneurship is gendered just as other aspects of society are, where women have to deal with problems of access and agency. Specifically, Indigenous entrepreneurs have a problem accessing finance to scale their businesses due to their exclusion from formal financial institutions. Most

Indigenous entrepreneurs are not deemed desirable by conventional and formal financial institutions, which impedes their ability to access credit or other financial services. The broader implication of this is that Indigenous entrepreneurs lack the requisite financial services or credit to expand and grow their businesses to be more economically viable. Despite this exclusion from the formal financial system and before this, Indigenous entrepreneurs developed their own means of financing and credit systems to help support their businesses. This financing system serves as an alternative to the formal banking system and is often based on a communitarian principle.

To further understand the intricacies of this system, this chapter engages Indigenous entrepreneurship and financing systems using Esusu as a case study. The chapter examines Esusu as an Indigenous financing system that Indigenous entrepreneurs have devised for their businesses in the face of their exclusion from formal financial services. It adopts a decolonial feminist lens to make sense of the exclusion of indigenous women entrepreneurs from the formal financial system and how Esusu is an agentic strategy that corrects this marginality. The chapter is divided into four substantive sections; the first section provides an overview of indigenous entrepreneurship and the discussions around it. The second section examines the financing for entrepreneurship in Africa and issues around access to finance; this section engages the nuances around women entrepreneurs' inaccessibility to finance or financial services. The third section examines the Esusu financial system, its features, and how it works for indigenous entrepreneurship. The fourth section engages the significance of the Esusu financial system for the sustenance of indigenous entrepreneurship in Africa.

Theoretical framework

Decolonial feminism is a theoretical lens developed by Lugones (2010). It centers a decolonial approach within a racial/gendered feminist context (Manning 2021). Decolonial feminist theory engages the discourse on coloniality/modernity and Global South indigenous identity and gender, while providing a space for the voices and experiences of silenced, "othered" women (Bhambra, 2014; Lugones, 2010). Applying Anibal Quijano's 1991 concept of coloniality within a gendered frame, Lugones stresses that the coloniality of gender continues to persist and that it lies at the intersection of gender/class/race as central constructs of the global capitalist system. She argues that the coloniality of gender enables the understanding of the oppressive imposition "as a complex interaction of economic, racialising, and gendering systems in which every person in the colonial encounter can be found as a live, historical, fully descriptive being" (Lugones 2010:747). She theorized decolonial feminism as a form of feminism that is "from and at the grassroots, and from and at the colonial difference, with a strong emphasis on ground, on a historicized, incarnate intersubjectivity" (Lugones 2010:746). It enacts a critique of racialized, colonial, and capitalist heterosexualist gender oppression as a lived transformation of the social. The work of non-Western, Indigenous, and women of colour involved in women's organizations inside their communities and society is the foundation of decolonial feminism (Lugones 2010). It suggests a new course that comes from marginalized women in the Global South and is not an imposition or prescription, but rather a way of thinking, acting, and comprehending gender.

Within the context of this chapter, decolonial feminism frames the understanding of the significance of Esusu for indigenous women entrepreneurs as operating outside the prescribed colonial line. Using this lens enables us to understand the otherness that Esusu represents outside of the formal financial institutions. Decolonial feminism allows us to think from the perspective of 'other' and align ourselves with the social and economic struggles of indigenous women entrepreneurs in Africa. This chapter engages the struggles of indigenous entrepreneurs vis-à-vis their exclusion from formal finance for their ventures. In doing so, decolonial feminism exposes us to the multiple oppressive systems that indigenous women entrepreneurs encounter, such as race, gender, and the economic system, and how these interact to manifest in their inability to access credit. Further, it helps us understand Esusu as a way that 'the other' circumvents their exclusion from formal finance by devising indigenous means of making sense of their world and affirming their agency. In the course of this chapter, this approach will be woven into the analysis to show how indigenous women entrepreneurs in Africa use Esusu to take charge of their agency and finance their business.

Understanding Indigenous Entrepreneurship

Entrepreneurship based on Indigenous knowledge and cultural expressions celebrates the rich and diverse cultures and traditions of Indigenous peoples and communities globally. This strengthens their sense of identity and belonging while also generating income and jobs (WIPO 2024). According to Tengeh et al. (2022), Indigenous people are constantly searching for methods to rebuild and enhance their communities or rescue them from the current state of depravity and poverty. Indigenous communities participate in a wide array of livelihoods and entrepreneurial endeavors, encompassing traditional hunter-gatherer practices and subsistence agriculture, as well as excelling as professionals within industrialized societies (Peredo et al., 2004).

Dana (1998) was among the first researchers to study indigenous entrepreneurship; he attributed the low level of entrepreneurial activity among native Peruvians to Western civilization's historical incorporation and distortion of their business ecosystem. Motivating the need to understand the entrepreneurial endeavors of Indigenous peoples within the context of their distinct individual and collective identities. The identities in question exhibit a rich diversity both within and across national boundaries; nonetheless, they collectively acknowledge their designation as Indigenous peoples. Indigenous entrepreneurship highlights the inherent diversity of Indigenous people and shows how the basic presumptions of mainstream theories might not accurately depict Indigenous people, their culture, their entrepreneurial style, and their growth path (Dana 2015). For instance, instead of assimilating into the capitalist economy, indigenous people would want to preserve their traditional economic structure (Dana and Anderson 2007). Indigenous communities prioritize kinship connections, with traditional knowledge disseminated through these intricate networks (Olaitan and Oloruntoba 2023). How Indigenous businesses establish relationships, function, and innovate may exhibit distinct characteristics. The rules and values that are ingrained in Indigenous civilizations have an impact on the character of entrepreneurship. Reavley et al. (2006) note that one of the main goals for Indigenous peoples is how commercial operations may respect and encourage the growth of Indigenous language and culture while also aiding in the development of a community. OECD (2019) notes that land assets are integral to Indigenous entrepreneurship, as most of these enterprises cover food, agriculture, mining, extractive industries, and residential and commercial real estate.

Dana and Anderson (2007) argue that the extent and character of innovation within Indigenous-owned enterprises are shaped by social structures rooted in kinship and collective cultural principles. For instance, emerging business initiatives are influenced by their cultural environment regarding the promotion of Indigenous languages and cultural traditions. In Indigenous communities, particularly in rural areas, the dynamics of innovation are likely to be influenced by the community's connection to the land (Drahos and Frankel 2012). Community-owned enterprises may also be operated by Indigenous groups; these enterprises can function as a subsidiary of a trust or as an independent legal entity under the governance of a designated body. They can assist in risk management by enabling the community to collectively absorb initial losses and cultivate a portfolio of enterprises to diversify the local economy and promote innovation.

Cornell (2006) examined the different motivations for encouraging Indigenous entrepreneurship. The first is that it can boost decision-making autonomy and decrease dependent ties, which helps to promote self-determination. In some countries, Indigenous Peoples have experienced higher rates of individual and collective forms of dependency, such as government programs and subsidies, and a lack of entrepreneurial activity as a result of colonization and assimilation processes (Cornell 2006). Through local job possibilities and revenue generation for public goods, such as services on traditional territories, Indigenous companies can assist in overcoming dependency (NSW Ombudsman 2016; Native Nations Institute, 2016; Cornell, 2006). The second is that Indigenous business leaders and entrepreneurs serve as significant role models for other Indigenous people. The third benefit is that it can support regional economic growth and preserve economic activity on traditional lands.

Indigenous peoples often establish development targets to meet social needs, provide communal benefits, and enhance community cohesion through entrepreneurship. In this context, social entrepreneurship serves as a mechanism to leverage entrepreneurial initiatives and market dynamics to attain these objectives. The principal aim of these enterprises is the enhancement of societal welfare rather than profit maximization (OECD 2019). The promotion of culture through Indigenous entrepreneurship is a vital component of sustainable development, as it cultivates an appreciation for Indigenous traditions and heritage. Indigenous entrepreneurship frequently engages the local community. Indigenous entrepreneurs can gain from experiential learning, which may include initiating and managing small enterprises, participating in community initiatives, and engaging with traditional customs. (Matlala et al. 2023).

Dana (2015) explains that, in contrast to modern technology-driven businesses that have a greater tendency to be wasteful and less environmentally sustainable, Indigenous entrepreneurship embodies sustainable practices. They are typically characterized by resource efficiency as they seek to obtain maximum value from limited resources (Dana and Anderson 2007). In addition to meeting local needs that are incomprehensible to outsiders, it promotes inclusivity, another tenet of sustainable development (Pansera and Sarkar 2016). Indigenous entrepreneurship is highly relevant not just to Indigenous peoples but to the larger society as well; this suggests that Indigenous entrepreneurship elevates inclusivity and sustainability. Compared to non-Indigenous entrepreneurship, Indigenous entrepreneurship is still growing at a slower rate. Maritz et al. (2022) note that this might be the result of obstacles like racial prejudice, a lack of education or training, a lack of business experience, or a lack of finances.

In addition to being a means of promoting the cultural underpinnings of Indigenous communities, entrepreneurship facilitates social, economic, and technological advancement. Indigenous entrepreneurs facilitate social innovation because they are deeply ingrained in social networks, family values, and other societal norms. Indigenous entrepreneurs, niche business promotion, and job prospects benefit from this (Maritz et al., 2022). They add that First Nations communities in Australia find Indigenous enterprises crucial (ibid para 1). By increasing workforce engagement, these companies can generate jobs for these communities. Indigenous populations profit greatly from this in terms of their health and quality of life, and the local economy also benefits. They point out that rather than focusing only on financial gain, some Indigenous business owners prioritize helping their communities. Nonetheless, there are still a lot of difficulties and barriers to overcome.

One of the main drivers of economic growth in Indigenous communities is the presence of indigenous entrepreneurs. They hire workers, build businesses, and create jobs. The fact that Indigenous businesses contribute to the preservation of their culture and identity makes them significant as well. Through the creation of jobs, resources, wealth, and inclusive opportunities, Indigenous enterprises contribute to the economic development of Indigenous communities. For Indigenous tribes, they are a source of pride and strength (Scaffeo 2022). Indigenous women view entrepreneurship as a means of improving the socioeconomic conditions in their community and the standard of living for their families. An important benefit of being an entrepreneur is having control over one's schedule, which facilitates the balancing of work and family obligations. In general, Indigenous peoples can use their entrepreneurial abilities to generate possibilities for their communities through Indigenous entrepreneurship. Indigenous entrepreneurship is significant because it empowers Indigenous people to drive economic growth within their communities. Additionally, it contributes to the preservation of their identity and culture (Scaffeo 2022).

Formal Financial Institutions and Access to Finance for Indigenous Entrepreneurs

Studies have examined the relationship between entrepreneurship, growth limitations, and restricted access to financial services faced by small businesses. Beck et al. (2005) studied the impact of financial limitations on the growth of small and medium enterprises (SMEs); they discovered that financial barriers are substantially and negatively correlated with the growth rate of enterprises, with small enterprises constantly suffering the most. SMEs routinely report more growth barriers than medium-sized or big businesses (Beck and Demirguc-Kunt 2006). Galindo and Schantiarelli (2003) argue that small businesses are more growth-constrained and have less access to financing than their large business counterparts in both developed and developing countries. Brixiova (2020) explains that entrepreneurship is hampered by limited access to credit and, consequently, investment capital, which can occur even in situations where the banking system has liquidity. Beck et al. (2005) show that SMEs' employment growth is positively impacted by easing financial limitations. The emphasis is on private SMEs, which, despite being a major employer, are the most economically marginalized. Although the private sector is thought to be responsible for 90% of jobs on the continent, research indicates that up to one-third of private companies in Africa cite restricted access to financing as a major restriction (McKinsey & Company 2012).

Indigenous entrepreneurs make up most of the African SME ecosystem, and they fall within the lower rung of the funding ladder. Due to limited financial resources, Africa's SMEs struggle to establish and eventually expand. Their primary sources of funding are their revenue, unofficial savings and loan organizations, which, due to their sectoral or regional specialization, are insecure and offer little opportunity for risk sharing. African SMEs have limited access to formal financing due to insufficient financial facilities and significant default risk. Financial institutions view SMEs as high-risk and, therefore, set difficult requirements before these SMEs can satisfy their financing requirements. In addition, few financial instruments are available in most of Africa due to the underdeveloped financial system. For instance, capital markets are still in their infancy, and SMEs cannot access long-term funding. The majority of financial systems in Africa are disjointed.

Modern technology makes financial services available in industrialized countries, but this technology is less developed and available in developing nations, particularly in rural areas with low levels of technological advancement (Kouladoum et al. 2022). Morrissette (2007) states that personal savings, family support, and borrowing from friends are the most common sources of funding for SMEs. While SMEs and startups depend on specific sources of funding, they often look for alternative external sources of funding because, in most cases, the money from these sources is never sufficient to support long-term organizational growth. Bank loans are one alternative, but they are rarely used because of the terms that financial institutions place on borrowing (Olarewaju et al. 2018). Most businesses find it difficult to obtain loans as they lack the necessary assets to serve as security for such loans. Wong et al. (2009) note that loans are a costly and cumbersome source of early-stage financial financing for new enterprises.

A gendered dimension exists between access to finance for women and male-led indigenous businesses. The growth trajectories of women-led firms differ from those of their male counterparts, as businesses run by women are typically smaller, less efficient in their production, less connected to institutions and markets, and have a harder time getting financing on the same conditions as men. Additionally, lack of access to money or financial services is a major limitation for micro, small, and medium enterprises (MSMEs) run or controlled by women. Consequently, women make up a disproportionate amount of the financing gap that indigenous businesses face, with the International Finance Corporation (IFC) estimating that the financing shortfall for women-owned MSMEs was USD 1.7 trillion in 2017. Also, compared to their male counterparts, women-led MSMEs are less active in foreign commerce; women only lead or own one out of every five exporting businesses. Given that businesses involved in trade are often more productive than those that are not, increasing women's engagement in trade by making it easier for them to obtain financing can improve competitiveness. Women exporters also have other obstacles to overcome; they frequently have restricted access to networks because of under-representation in leadership roles within networks that are controlled by men. Other factors that disproportionately hinder their involvement in export commerce include a lack of digital skills, a lack of resources, complicated bureaucratic procedures, high compliance costs, ambiguous legal frameworks, and a lack of knowledge about regional, continental, and international trade agreements (Malunga 2024).

Specifically, women are most affected by the financial difficulty that indigenous entrepreneurs face. There is a consensus that an important factor and powerful force behind inclusive growth is financial inclusion, particularly women's financial inclusion. Financial services can increase women's employability and labour by providing funds for training and education (Asongu and Odhiambo 2018; Asongu et al. 2020). However, women entrepreneurs face unique challenges; for example, women make up 56% of the adult population that is still "unbankable", that is, without access to a bank account, with a nine-percentage point gender disparity, based on the World Bank's 2017 Global Findex Report. Morsy and El-Shal (2019) report that in Africa, 48 percent of men and 37 percent of women hold bank accounts, a difference that has grown over the previous few years. Also, women face disproportionate barriers to financial resources, such as high loan rates and complicated collateral requirements. Due to things like insufficient financial knowledge, risk aversion, and failure-related anxiety, women entrepreneurs in the credit market do not even apply for loans (Morsy 2020). Kou^baa (2014) explains that when women entrepreneurs apply for a bank loan, they typically face gender biases and strict credit application procedures. To address some of the difficulties that Indigenous women entrepreneurs face, banks or microfinance institutions can supply productive microfinance, which is necessary and goes hand in hand with payments.

Better knowledge of how to create financial sector ecosystems that encompass more than a few providers and can offer low-income women the products they require is required. There needs to be more government involvement, cheaper wholesale financing in the local currency, stronger and more digital financial service providers, and better market infrastructure, especially in the data space (Delling 2024).

Embracing a decolonial lens enables the understanding of the invisibilization of indigenous entrepreneurs from the formal financial system as a consequence of coloniality and its ability to otherise people into beings and non-beings. Quijano (1991) theorized coloniality to explain how the colonial system creates a bifurcation of people into beings and non-beings using the three colonialities (of power, of being, and of knowledge). Specifically for indigenous women entrepreneurs, decolonial feminism engages the coloniality of gender as central to understanding the workings of the colonial system and its effects on the colonized along gender lines. This lens moulds our understanding of why indigenous entrepreneurs are faced with different hurdles in business and why indigenous women entrepreneurs are less likely to access credit. Additionally, the global capitalist system that created the formal financial institutions was conceptualized to cater to the colonial interests. Scholars engaging the coloniality/modernity discourse engage how the colonized/global south were integrated into this system at an unequal level, leaving them in a disadvantaged position. This forced integration into the capitalist system renders the global south unable to maximize the benefit of the system due to their lack of economic power and the interplay of systems of power such as race, economic, social, and gender. These systems of power make it inherently difficult for indigenous entrepreneurs from the global south to fairly navigate the capitalist system, with an even worse scenario for indigenous women entrepreneurs by virtue of the coloniality of gender.

Financing Indigenous Entrepreneurship Through Esusu

The exclusion of indigenous entrepreneurs from formal financial institutions has consolidated their use of alternative financing means, such as informal financing, community savings groups, most of which are rooted in indigenous knowledge and practices. Kauffman (2005) notes that microcredit institutions, whose expansion is attributable to the flexible loans they provide to small enterprises, have partially compensated for the lack of finance for SMEs. Informal financing is another tactic that women entrepreneurs are using to combat inequality in the financial system. Richardson et al. (2004) note that women entrepreneurs are more likely than men to finance their businesses through informal financing methods, based on a survey of small businesses in three African nations: Ethiopia, Tanzania, and Zambia (cited in Morsy and El-Shal 2019). Among these unofficial funding sources is the Esusu system, which is unique to the Yoruba people of southwest Nigeria.

Esusu is a customary method of collaboration among the Yoruba people of Nigeria, whereby groups of people make contributions to credit and savings organizations for their mutual gain. Bascom (1952) notes that the Esusu is a fund to which a group of people donates a set amount of money at predetermined periods; the group's total contribution is then divided among the members in turn. These organizations are primarily involved in credit financing and agricultural production, which complements and replaces contemporary official financial systems and cooperative institutions (Osabuohien and Ola-Davi 2018). The Esusu system is considered an old financing institution by the Yoruba people; it is thought to have started among Nigeria's Yoruba people and then spread to Liberia, the Democratic Republic of the Congo, and the majority of West African countries (Seibel 2004). Bascom (1952) explains that this financing system is indigenous to the Yoruba people, and there is not a single European organization that can be thought of as a model from which the Esusu might have originated, especially when it comes to those documented among Europeans in West Africa.

Research on informal finance frequently links Esusu to revolving credit and savings organizations (Seibel 2004). The term Esusu refers to the fund and not the group of contributors (Bascom 1952:63). In Esusu, people band together using informal finance methods to advance their individual and group financial goals. There are various ways that the Esusu manifests; some organizations mobilize savings but do little to no lending. Second, some lending organizations use minimal savings mobilization. Third, some organizations practice self-help finance and use different kinds of savings, such as those offered by licensed cooperatives and rotating ones (Ojenike and Olowoniyi 2013).

In terms of the operation, a group of people bands together to contribute a set and equal amount of money at predetermined intervals, such as daily, weekly, fortnightly, monthly, or bimonthly, so that each member can turn in the full amount in turn. The entire group utilizes the pool before initiating a new cycle of rotation. Typically, a ballot or consensus determines the sequence of money drawing. In cities, the Yorubas make a distinction between Ajo and Esusu. In Ajo, a professional collector known as the Alajo collects donations, typically on a daily or weekly basis, in exchange for a small commission. The Esusu and Ajo systems are more intimate than the formal financial system, as they allow deposits without the need for physical bank visits. Although Ajo contributors may not know one another, it is less intimate than the Esusu system. Instead, the Alajo acts as a mediator between them, recording donations. While they are free to donate in accordance with their budget, each contributor is required to make consistent contributions within a specified time frame. Ajo differs from Esusu in that a donor may request that the Alajo refund a specific portion of their contribution in specific situations, such as an emergency (Osabuhien and Ola-David 2018). In contrast to other informal financing systems, many Esusu groups do not meet, and their members are often unfamiliar with one another. Even the leader of the Esusu organization may be unaware of the members' identities. Each group differs in the number of donors, the quantity of the contributions, and the duration of the intervals between them. In theory, each member receives a single lump sum payment equal to the amount of contributions they made for a single cycle. In the Esusu system, the group leaders typically do not make a profit; the only benefits they enjoy are the recognition as the group's leader and the gratification of coordinating the group by either accepting or rejecting their funding requests. Fadipe (1970) explains that the Esusu leader receives the entire sum contributed, together with a portion that is typically set aside for administrative purposes.

Although participation in Esusu is typically voluntary, the group influence is very present, and savings are regularly and inevitably required. Despite being optional, each member is required to make their contribution when it is due (Fadipe 1970; Bascom 1952). Generally speaking, defaults or late contributions are frowned upon and punished severely. In certain circumstances, members may have their request for an early turn rejected or be expelled outright (Olarewaju et al. 2018).

Esusu relies on trust and loyalty, considering that it operates outside of conventional and formal financial institutions and systems. This prevents early-collection association members from withdrawing from the system, which would result in other members losing some or all of their contributions. Its popularity remained high even after the establishment of formal microfinance banks in Nigeria (Osabuhien and Ola-David 2018). Religious organizations, market areas, rural and urban communities, and workers in the unorganized sector all use it. People with low and moderate incomes and rural workers especially prefer it to a formal financial system because they are underpaid and hence unable to use the official banking system. Compared to the official banking system, Esusu offers more convenient saving options and lower credit costs because it customizes its financial services to meet the actual, daily requirements of every group member (Oloyede 2008). Two benefits that Esusu offers in comparison to the conventional financial system are that it is interest-free and also a form of gendered financing. Joining an Esusu club is interest-free, which is a major cost-related inducement for Indigenous entrepreneurs. On gender, Esusu is a more informal way of saving money and accessing credit for women entrepreneurs (National Bureau of Statistics 2013). Although Esusu is primarily an informal practice, it is prevalent in formal work environments and provides benefits to both individuals and groups within corporate organizations. It is common to view Esusu as an adjunct to other credit-obtaining methods (like cooperatives). A lot of people use Esusu to work towards a certain goal, like buying property, launching a company, or growing their business. Informal company owners frequently use Esusu when they struggle to secure bank loans (Osabuhien and Ola-David 2018).

Allen et al. (2019) frame Esusu as a rotating savings group (RSG) in which members pool and share financial resources. They define an RSG as a community funding program for financial resources provided by friends, relatives, and neighbors, as well as group-based loans and savings wherein members of a group contribute a certain amount each month, and one member of the group collects the entire money raised that month on a rotating basis (Bruton et al., 2021). Such a fund-raising technique makes up a far larger percentage of transactions in many developing countries, enabling their underprivileged yet entrepreneurial communities to acquire start-up capital or enhance productive assets like family farms (Chen et al. 2017).

Although Indigenous entrepreneurs have found Esusu and other informal financing systems beneficial, other research downplays their importance as a source of funding for businesses. According to Gu'erin et al. (2011), informal finance is considered bad finance, while De Rambures et al. (2017) characterize it as a remnant of outdated lending processes that modernization will eventually replace. Degryse and Ongena (2016) describe informal finance as underground financing, which suggests that it involves criminal behavior.

Significance of the Esusu Financing System for Indigenous Entrepreneurship

Moses et al. (2015) examine the benefit of Esusu for indigenous entrepreneurship and find a positive correlation between the Esusu scheme and the achievement of entrepreneurial goals by indigenous entrepreneurs. They discovered that the participation of entrepreneurs in the Esusu scheme significantly impacts their ability to achieve entrepreneurial goals. This shows that the Esusu system can finance indigenous entrepreneurship.

Simba et al. (2023) assert that community financing programs like Esusu gave a significant portion of women entrepreneurs an alternative means of getting start-up funding. They reported that most women entrepreneurs stated that they used the money they received from the program as startup capital for their businesses. They also acknowledged that they depended on the plan to provide them with the money they needed for their regular business ventures since it ensured them income and kept their enterprise viable. Their justification for choosing the program is that it offered advantageous terms that they might not have had when applying for a bank loan (Beck & Demirgüç-unt, 2006). Brunton et al. (2011) and Chliova et al. (2015) add that community funding programs help women escape poverty by giving them access to entrepreneurship and promoting the economic development of their respective regions.

The integrity and confidence of the contributing members in Nigeria form the foundation of Esusu finance schemes (Hofstede 1980). Their sense of community and belonging seemed to foster a certain level of trust, which is crucial to the operation of the program. Based on the aforementioned statements, it is clear that the RSG program promoted tight collaboration and community involvement among women entrepreneurs. Most women entrepreneurs agreed that the sense of community that these programs fostered allowed them to overcome their problems by feeling connected to one another. In other words, these programs united their business community. Simba et al. (2023) report that through the community financing program, there was a sense of solidarity, which improved their communal ties, thereby allowing them to access shared community resources, such as funding for investments connected to entrepreneurship. They note further that participation in community-based initiatives promotes collectivism, which serves as the foundation for group cohesion and solidarity in developing countries. Such circumstances provide advantageous communal resources for women entrepreneurs, who frequently face discrimination due to their gender. The plan served as a community bridge, fostering ties and solidarity among these women (Burt 2019).

Community financing programs serve as a means of supporting women entrepreneurs in developing countries. Bhatt and Tang (1998) argue that such a community-based funding mechanism might be viewed as a supplementary program or scheme that operates in conjunction with regulated microlending financial services to provide a solution for women entrepreneurs in developing economies. Programs like Esusu provide a variety of purposes, including enterprise financing, consumption smoothing, encouraging savings discipline, and mediating transactions between savers and borrowers. Instead of depending on anonymous interactions between a consumer and a formal institution, informal financial institutions are arguably distinguished by their emphasis on interpersonal ties (Aliber 2015).

Most formal financial institutions are unable to implement the transaction cost reduction techniques that informal financial institutions, such as Esusu, can use to attract entrepreneurs (Aliber 2015). Additionally, researchers found that Esusu groups have significantly contributed to the socioeconomic and infrastructure development of their community. The members' use of their donations for impactful endeavors like farming, entrepreneurship, and ward sponsorship may account for this. For many Indigenous entrepreneurs, Esusu groups are their last resort due to a number of disappointments they have encountered when interacting with banking institutions, including the inability to obtain loans, the issue of wasting time when taking out their money, and the need to document every transaction whenever they interact with the banks (Thomas 2015).

The significance of Esusu for indigenous entrepreneurs and indigenous women entrepreneurs goes beyond the provision of credit. A decolonial approach shows us how Esusu represents a shift from the Euromodern system of finance and a challenge to the global capitalist system. Esusu, as an institution, erases the exclusion and marginality encountered by indigenous entrepreneurs within formal financial institutions due to modernity/coloniality by providing an alternative means of entrepreneurship finance. It is a financing scheme that contests the bifurcation created by the global capitalist system and formal financial institutions by opening the space for indigenous entrepreneurs and specifically indigenous women entrepreneurs. In a way, it is an agentic strategy that indigenous entrepreneurs use to affirm their agency as entrepreneurs, while contesting the exclusion and marginality they experience within the formal financial system. Esusu, therefore, becomes an important manifestation of a decolonial approach to entrepreneurship financing and a form of decolonial feminist organizing. The latter becomes obvious by centering how indigenous women entrepreneurs can build community, experience belonging, and solidarity within their Esusu group. Esusu holds much significance for indigenous entrepreneurs beyond access to credit, as it helps correct the inequality produced by the racialized and global capitalist system. It offers an alternative credit system for indigenous entrepreneurs rooted in indigenous knowledge and practice that can be adopted across different contexts. It encapsulates the persistent call for a decolonial approach to entrepreneurship financing in Africa.

Conclusion

This chapter examined the financing system developed by indigenous entrepreneurs to finance and expand their businesses. Specifically, it examined the Esusu financial system as one of the alternative means of finance for indigenous entrepreneurship in Africa. The chapter started by providing an understanding of Indigenous entrepreneurship; it notes that Indigenous entrepreneurs and their businesses are drivers of social innovation because they are often embedded in family, social values, and networks. In this chapter, it was argued further that Indigenous entrepreneurship is a means for

Indigenous people to maintain their traditional economic system rather than integrate into the modern capitalist economy. The chapter also examined the financial system for entrepreneurship in Africa and access to finance for indigenous entrepreneurs. It noted that women entrepreneurs have difficulty accessing financing services due to the elitism that the formal financial system espouses. In addition, the many requirements of this system make it difficult for indigenous entrepreneurs to access credit, loans, and other financial services. The chapter highlighted the exclusion of Indigenous entrepreneurs from the formal financial system as one of the motivations for the alternative financial system they built. Although most of these alternative financial systems predate formal banking systems, Indigenous communities have used them to finance their businesses. Taking all this into consideration, it is evident that we cannot begin to discuss how Africa can harness the potential of entrepreneurship for development if there are intricate challenges that need to be addressed. Entrepreneurship is yet to be duly recognized as an important element of Africa's development, explaining why entrepreneurs on the continent still face challenges. Women entrepreneurs face challenges of financing, access to credit, and engagement with formal financial systems. Considering that most women entrepreneurs are also indigenous entrepreneurs, the discourse becomes multi-layered. There have been commitments by governments and international organizations to democratize access to finance for indigenous entrepreneurs in Africa, but this has yet to manifest holistically. Specifically, Indigenous entrepreneurs are often removed from relevant conversations around funding, financial services, entrepreneurship policy, etc. The exclusion of Indigenous entrepreneurs from the formal financial system consolidated the need for an alternative financing system to serve their businesses.

Esusu, which is the major focus of this chapter, is one of the alternative systems that Indigenous entrepreneurs have devised to provide financing for their businesses. It represents a form of communal funding for businesses and does not require lots of elitist financial hurdles as the formal financial system does. It is argued that although Esusu is particular to the Yoruba people of Nigeria, we can find a semblance of this communal funding system in other cultures, for instance, *stokvels* in South Africa. Stokvels are an informal self-help union of people with a common goal of pooling funds together on a weekly or monthly basis (Bophela and Khumalo 2019:27). They operate within the informal sector and "are a platform for periodic collective savings for members" with mostly black South Africans in the townships, villages and cities practicing them (Arko-Achemfour 2012:129; Bophela and Khumalo 2019:27). The operation of stokvels exist outside of the formal financial system, just like Esusu and they create a pathway for members to access credit outside of the restrictive formal financial system. The informality attached to stokvels ties into the decoloniality discourse on the bifurcation between beings and non-beings, and the attachment of informality to indigenous practices outside the global financial system. Thereby creating another bifurcation with respect to financing, i.e., formal and informal financial institutions. The chapter recommends the need to embrace an indigenous financing system for entrepreneurs to circumvent the problems of access to formal financial institutions in Africa. Esusu, just like stokvels in South Africa and other parts of Africa, provides a getaway for some of the entrepreneurship woes that Africa suffers today. For instance, access to finance continues to hinder the maximization of entrepreneurship for socioeconomic development on the continent. This chapter, therefore, engages how Esusu can address this problem and, on a more nuanced level, kickstart the plan for devising a decolonial approach to socioeconomic development in Africa. Considering that Esusu transcends the racial, colonial, and gendered lines drawn by colonial structures, it sets the path for framing what a decolonial financial system looks like in Africa and how this system can correct the marginality created by coloniality/modernity.

References

Aliber, M. (2015). *The importance of informal finance in promoting decent work among informal operators: A comparative study of Uganda and India* (Working Paper No. 66). International Labour Office.

Arko-Achemfuor, A. (2012). Financing small, medium and micro-enterprises (SMMEs) in rural South Africa: An exploratory study of stokvels in the Naledi Local Municipality, North West Province. *Journal of Sociology and Social Anthropology, 3*(2), 127–133.

Asongu, S. A., & Odhiambo, N. M. (2018). ICT, financial access, and gender inclusion in the formal economic sector: Evidence from Africa. *African Finance Journal, 20*(2), 46–66.

Bascom, W. R. (1952). The Esusu: A credit institution of the Yoruba. *The Journal of the Royal Anthropological Institute of Great Britain and Ireland, 82*(1), 63–69.

Beck, T., & Demirgüç-Kunt, A. (2006). Small and medium-size enterprises: Access to finance as a growth constraint. *Journal of Banking & Finance, 30*(11), 2931–2943.

Beck, T., Demirgüç-Kunt, A., & Maksimovic, V. (2005). Financial constraints to firm growth: Does firm size matter? *The Journal of Finance, 60*, 137–177.

Bhambra, G. K. (2014). Postcolonial and decolonial dialogues. *Postcolonial Studies, 17*(2), 115–121.

Bhatt, N., & Tang, S. Y. (1998). The problem of transaction costs in group-based microlending: An institutional perspective. *World Development, 26*(4), 623–637.

Bophela, M., & Khumalo, N. (2019). The role of stokvels in South Africa: A case of economic transformation of a municipality. *Problems and Perspectives in Management, 17*(4), 26–37.

Brixiová, Z., Kangoye, T., & Yogo, T. U. (2020). *Access to finance among small and medium-sized enterprises and job creation in Africa* (GLO Discussion Paper No. 665). Global Labor Organization.

Bruton, G. D., Nuhu, N., & Qian, J. J. (2021). Informal finance in settings of poverty: Establishing an agenda for future entrepreneurship research. *Journal of Developmental Entrepreneurship, 26*(2), Article 2150011.

Burt, R. S. (2019). Network disadvantaged entrepreneurs: Density, hierarchy, and success in China and the West. *Entrepreneurship Theory and Practice, 43*(1), 19–50.

Chen, J., Chang, A. Y., & Bruton, G. D. (2017). Microfinance: Where are we today, and where should the research go in the future? *International Small Business Journal, 35*(7), 793–802.

Chinonye, M. L., Akinbode, M., & Obigbemi, I. F. (2015). Indigenous financing: An effective strategy for achieving the entrepreneurial objectives of SSEs. *Journal of Entrepreneurship: Research & Practice, 2015.*

Chliova, M., Brinckmann, J., & Rosenbusch, N. (2015). Is microcredit a blessing for the poor? A meta-analysis examining development outcomes and contextual considerations. *Journal of Business Venturing, 30*(3), 467–487.

Cornell, S. (2006). *What makes First Nations enterprises successful? Lessons from the Harvard Project.* http://www.ksg.harvard.edu/hpaied

Dana, L. P. (1988). International note: More small business is not the answer for Peru. *Journal of Small Business Management, 26*(3), 68–70.

Dana, L. P. (2015). Indigenous entrepreneurship: An emerging field of research. *International Journal of Business and Globalisation, 14*(2), 158–169.

Dana, L. P., & Anderson, R. B. (2007). Taos Pueblo: An Indigenous community holding on to Promethean values. *Journal of Enterprising Communities: People and Places in the Global Economy, 1*(4), 321–336.

Degryse, H., Lu, L., & Ongena, S. (2016). Informal or formal financing? Evidence on the co-funding of Chinese firms. *Journal of Financial Intermediation, 27*, 31–50.

De Rambures, D., & Dueñas, F. E. (2017). Credit black market. In *China's financial system: Growth and inefficiency. Palgrave Macmillan.* https://doi.org/10.1007/978-3-319-40451-6

Delling, D. (2024, April 16). Credit where it's due: How better access to affordable capital for women can unlock growth for all. *Mastercard Foundation Newsroom.* https://newsroom.mastercard.com/news/perspectives/2024/credit-where-it-s-due-how-better-access-to-affordable-capital-for-women-can-unlock-growth-for-all/

Drahos, P., & Frankel, S. (2012). *Indigenous people's innovation: Intellectual property pathways to development.* ANU E Press.

Fadipe, N. A. (1970). *The sociology of the Yoruba.* University of Ibadan Press.

International Finance Corporation. (2017). *MSME finance gap: Assessment of the shortfalls and opportunities in financing micro, small and medium enterprises in emerging markets.* IFC.

Kauffmann, C. (2005). *Financing small and medium enterprises in Africa* (Policy Insights No. 7). OECD Development Centre.

Koubaa, H. F. (2014). Start-up informal finance and formal microfinance: The role of human capital. *Strategic Change, 23*(7–8), 415–423.

Kouladoum, J.-C., Wirajing, M. A. K., & Nchofoung, T. N. (2022). Digital technologies and financial inclusion in Sub-Saharan Africa. *Telecommunications Policy.* https://doi.org/10.1016/j.telpol.2022.102387

Lugones, M. (2010). Toward a decolonial feminism. *Hypatia, 25*(4), 742–759.

Malunga, M. F. (2024, April 7). Accelerating access to finance for women entrepreneurs: Gateway to regional, continental, and international trade. *tralac Blog.* https://www.tralac.org/blog/article/16374-accelerating-access-to-finance-for-women-entrepreneurs-gateway-to-regional-continental-and-international-trade.html

Manning, J. (2021). Decolonial feminist theory: Embracing the gendered colonial difference in management and organisation studies. *Gender, Work & Organization*, *28*, 1203–1219.

McKinsey & Company. (2012). *Africa at work: Job creation and inclusive growth.* McKinsey Global Institute.

Morrissette, S. (2007). A profile of angel investors. *The Journal of Private Equity, 10*(3), 52–66.

Manning, J. 2021. Decolonial feminist theory: embracing the gendered colonial difference in management and organisation studies. Gender Work Organisation, *28*:1203–1219.

Maritz, A., Andrews, J., Jones, C., and Foley, D. 2022. Indigenous entrepreneurship key. 22 February, LaTrobe University. https://www.latrobe.edu.au/news/articles/2022/release/indigenous-entrepreneurship-key2

Matlala, N., Sindi, B., and Tsengwa, R. 2023. Indigenous Entrepreneurship Education and Sustainable Development Key for Economic Growth. 10 November, University of Western Cape. https://www.uwc.ac.za/news-and-announcements/news/indigenous-entrepreneurship-education-and-sustainable-development-key-for-economic-growth

Morsy, H. (2020, March). Access to finance: Why aren't women leaning in? *Finance & Development.* International Monetary Fund. https://www.imf.org/en/Publications/fandd/issues/2020/03/africa-gender-gap-access-to-finance-morsy

Morsy, H., El-Shal, A., & Woldemichael, A. (2019). Women self-selection out of the credit market in Africa (Working Paper No. 317). African Development Bank Group.

National Bureau of Statistics. (2013). Use of financial services in Nigeria. NBS & World Bank.

Native Nations Institute. (2016). *Access to capital and credit in Native communities.* http://nni.arizona.edu

NSW Ombudsman. (2016). *Community and disability services: Fostering economic development for Aboriginal people in NSW.* NSW Ombudsman.

Organisation for Economic Co-operation and Development. (2019). Promoting Indigenous entrepreneurship and small business development. In Linking Indigenous communities with regional development. OECD Publishing.

Ojenike, J. O., & Olowoniyi, A. O. (2013). Estimating growth in investment of micro and small scale enterprises in Nigeria. *Asian Economic and Financial Review, 3*(1), 111–123.

Olaitan, Z. M. (2024). The relevance of African Indigenous knowledge systems for development in Africa. AIKRN Policy Brief Series.

Olaitan, Z. M., & Oloruntoba, S. O. (2023). Beyond reaction: (Re)-imagining African agency in the decolonization of knowledge. In N. Andrews & N. Khalema (Eds.), *Decolonizing African pedagogies: Knowledge production, epistemic imperialism, and Black agency* (pp. 31–53). Springer.

Olarewaju, A. D., Adebisi, S. A., & George, O. J. (2018). Indigenous African financing strategies as a unique engine for entrepreneurship growth. In Indigenous management practices in Africa (Vol. 20, pp. 147–166). Emerald Publishing. https://doi.org/10.1108/S1877-636120180000020008

Oloyede, J. A. (2008). Informal financial sector, savings mobilization and rural development in Nigeria: Further evidence from Ekiti State of Nigeria. *African Economic and Business Review, 6*(1), 35–63.

Osabuohien, E., & Ola-David, O. (2018). Esusu (Nigeria). In A. Ledeneva (Ed.), *The global encyclopedia of informality.* UCL Press.

Pansera, M., & Sarkar, S. (2016). Crafting sustainable development solutions: Frugal innovations of grassroots entrepreneurs. *Sustainability, 8*(1), Article 51.

Peredo, A. M., Anderson, R. B., Galbraith, C. S., Benson, H., & Dana, L. P. (2004). Towards a theory of Indigenous entrepreneurship. *International Journal of Entrepreneurship and Small Business, 1*(1), 1–20.

Quijano, A. (1991). Colonialidad, modernidad/racialidad. *Perú Indígena, 13*(29), 11–29.

Richardson, P., Howarth, R., & Finnegan, G. (2004). *The challenges of growing small businesses: Insights from women entrepreneurs in Africa.* International Labour Office.

Scaffeo, A. (2022, October 29). Why is Indigenous entrepreneurship important for our communities? *Digital Vibes.* https://www.digitalvibes.ai/post/why-is-indigenous-entrepreneurship-important-for-our-communities

Seibel, H. D. (2004). Microfinance in Nigeria: Origins, options, and opportunities. *Journal of Social Sciences, 17*(1), 63–71.

Simba, A., Ogundana, O. M., Braune, E., & Dana, L. P. (2023). Community financing in entrepreneurship: A focus on women entrepreneurs in the developing world. *Journal of Business Research, 163,* Article 113962.

Tengeh, R. K., Ojugebele, H., & Ogunlela, O. (2022). Towards a theory of Indigenous entrepreneurship: A classic? *International Journal of Entrepreneurship and Small Business, 45*(1), 1–15.

Thomas, A. (2015). Revolving loan scheme (Esusu): A substitute to the Nigerian commercial banking system? *Journal of Business and Management, 17*(6), 62–67.

World Intellectual Property Organization. (2024). *Indigenous and local community entrepreneurship.* https://www.wipo.int/tk/en/entrepreneurship/index.html

Wong, A., Bhatia, M., & Freeman, Z. (2009). Angel finance: The other venture capital. *Strategic Change, 18*(7–8), 221–230.

World Bank. (2017). *SME finance.* https://openknowledge.worldbank.org/handle/10986/28860

Chapter 6

INDIGENOUS WOMEN'S ENTREPRENEURSHIP IN AFRICA

Introduction

Currently, widespread poverty, inequality, and extreme hardship continue to affect millions of Africans (Thorbecke et al., 2024). This broader reality is mirrored in South Africa, where slow economic growth and persistently high unemployment have driven the economy into a precarious state (Mulibana & Tshikovhi, 2024). This economic turmoil derails progress toward achieving inclusive socioeconomic development (Croce, 2020). Women who experience higher unemployment rates (Posel, 2025) are disproportionately affected by unemployment and poverty (Majokweni et al., 2024). Hence, women's entrepreneurship is not only an economic imperative but also a powerful vehicle for advancing socioeconomic equality (Ebewo et al., 2025). However, the persistence of racial and gender hierarchies, rooted in Eurocentric ideologies, continues to reproduce inequalities in entrepreneurship, often silencing women's voices and limiting their access to funding, resources, and decision-making power. In this context, Indigenous knowledge remains an underutilized yet invaluable resource for fostering self-sufficiency and economic resilience, particularly within Indigenous communities (Tshamano et al., 2023). Indigenous Women's Entrepreneurship (IWE) embodies a reclamation of agency and a revitalization of ancestral knowledge. By leveraging their epistemological heritage, women challenge hegemonic models imposed by colonial and capitalist frameworks. This enhances self-determination, safeguards traditions, and fosters economic independence (Padilla-Meléndez *et al.*, 2022). It also highlights the need for a nuanced analysis at the intersection of Indigenous Entrepreneurship (IE) and gender to redress deep-rooted inequalities and foster entrepreneurial transformation.

While research on IE has grown (for example, see AlMehrzi et al., 2024; Colbourne, 2017; Demong et al., 2023; Selelo, 2023), the discourse remains devoid mainly of diverse gender perspectives, particularly those centering on Indigenous women (Croce, 2020). As a result, there is limited understanding of Indigenous women entrepreneurs' experiences. It is known, however, that Indigenous women endure multiple layers of systemic discrimination and marginalization, from the broader society but also within their own communities. As a result, their economic contributions and entrepreneurial roles remain underexplored, mainly due to the persistence of patriarchal colonial structures that continue to exclude them from decision-making spaces (Croce, 2020). Moreover, their voices continue to be silenced in academic discourse due to the intersecting effects of racism, sexism, and cultural discrimination (Menzies et al., 2024). Hence, the dominance of Eurocentric, masculinized entrepreneurial paradigms and narratives also sustains the illusion of a universal feminist experience that fails to account for distinct realities faced by Indigenous women entrepreneurs.

Conventional entrepreneurship reflects enduring colonial and socio-political inequalities. However, dominant narratives often strip this model of its political context, overlooking the historical and material influences of colonialism, racism, imperialism, and globalization. However, these forces have shaped access to entrepreneurial opportunities, determined which forms of enterprise are recognized, and dictated whose knowledge is considered valid. The main argument advanced here is that Indigenous women have been systematically marginalized, and their voices silenced or distorted within dominant entrepreneurial narratives. As such, the discourse on entrepreneurship cannot be divorced from the historical and contemporary context in which these women operate. That is, many women navigate entrepreneurial spaces shaped by colonial legacies, patriarchy, and land dispossession (Lugones, 2010). These historical conditions continue to shape the present, where women face persistent barriers such as limited access to finance, the burden of unpaid care work, and the pressures of neoliberal policy environments, among other barriers (Demong et al., 2023; Donga & Chimucheka, 2024). Recent scholarship has begun to explore IWE to fill this gap by centering their identities, challenges, and agency. For example, Croce's (2020) theoretical paper argues that recognizing Indigenous women entrepreneurs' identities and challenges fosters meaningful and sustainable empowerment. In Sarawak, Malaysia, Yap et al. (2023) found that rural Indigenous women entrepreneurs were primarily motivated by internal factors like personal ambition, financial independence, and a desire to contribute to household income. Moreover, their success was shaped by personal traits such as perseverance, self-belief, commitment, and risk-taking. Additionally, external factors, including social networks, financial support, business training, and favorable government policies, were crucial to their entrepreneurial success. Similarly, research by Menzies et al. (2024) highlights how Australian Indigenous women entrepreneurs draw strength and resilience from their ethnicity, gender, and Indigenous identity to enhance their capacity to navigate complex challenges.

However, despite these valuable insights, a gap remains regarding IWE within the South African context. This chapter addresses this gap by examining the intersection of women's Indigenous Knowledge (IK) and entrepreneurship in KwaZulu-Natal (KZN) through a decolonial lens to uncover the colonial underpinnings of entrepreneurship while illuminating the role and relevance of Indigenous Knowledge Systems (IKS) in entrepreneurship. To achieve this, it draws on existing literature on IE and incorporates insights gained from interviews with Indigenous women entrepreneurs from two communities in rural KZN: Ezinqoleni (Nguni) and Bergville (Abathwa).

A qualitative approach was employed to bring to the fore the "ontological reality of intersubjective knowledge", including the "epistemological and methodological implications of this reality" (Adler 1997:323). This is done to understand entrepreneurship through Indigenous women's perspectives rather than imposing universal theoretical constructs, recognizing that human experiences are fluid rather than fixed. Moreover, qualitative methods are appropriate since the study is exploratory, seeking depth rather than breadth (Powoh, 2016). To this end, semi-structured interviews were conducted to capture first-hand accounts of women's entrepreneurial activities. Indigenous women entrepreneurs were purposively selected from these communities based on their involvement in indigenous entrepreneurial activities. A total of 11 women were interviewed, which aligns with Creswell and Creswell's (2005) recommendation that qualitative research should include between five and 25 participants, as the focus is on detailed and context-specific insights rather than statistical generalization. The data were analyzed using thematic analysis, which allowed for the exploration of multiple perspectives, noting similarities and differences, while discovering new insights (Braun & Clarke, 2006).

This study elevates voices about IK and practices by, with, and for women entrepreneurs in selected Indigenous communities in KZN. By centering Indigenous women's voices, the study challenges dominant epistemologies and promotes alternative entrepreneurship frameworks that emerge from Indigenous ways of knowing and doing. Ultimately, this study aims to enhance Indigenous women's entrepreneurial capacities, promote gender equity, and advance sustainable development within Indigenous communities. Without such nuanced analyses, entrepreneurship policies and programs risk being oversimplified, ineffective, and disconnected from Indigenous women's lived realities. The rest of the chapter is structured as follows: The next section presents decolonial feminism as an analytical lens for this contribution. This is followed by a literature review, defining IE and IWE and situating both within existing scholarship. The subsequent sections present women's entrepreneurial activities, their motivations, and challenges. The chapter concludes by offering reflections and practical recommendations to strengthen support mechanisms for IWE in KZN.

On Decolonial Feminism

The discussion in this chapter is framed within decolonial feminism. Lugones (2010) introduced decolonial feminism as a theoretical lens to challenge the 'coloniality of gender,' which includes racialization, colonization, capitalist exploitation, and heteronormativity. Lugones (2010) critiques the idea of gender as a universal category, arguing that colonial powers imposed it to categorize and dominate colonized populations (Nassiri-Ansari & Rhule, 2024). Decolonial feminism thus explores identity, power, and resistance in postcolonial contexts, confronting Eurocentric, white, and Western epistemologies by affirming Indigenous and gender-expansive forms of existence (Lugones, 2010). That is, decolonial feminists seek to identify the inherent inequalities and also actively work to deconstruct the racist, capitalist, cisnormative, and heteropatriarchal ontological foundations of the world (Lugones, 2010). By centering marginalized women's lived realities, this approach builds empathy and solidarity with historically othered and silenced groups (Cele, 2023). Decolonial feminism, therefore, aligns with broader objectives of social and environmental justice, the rights of Indigenous people, and the decolonization of knowledge (Mignolo, 2007) by integrating diverse epistemologies and ontologies. It advocates shifting away from epistemic violence toward embracing a pluriverse of diverse knowledges, ideas, and experiences.

This is important because it needs to be remembered that in Africa, colonial powers imposed racial hierarchies that subordinated Indigenous communities and enforced Eurocentric ideologies on Indigenous lands. These hierarchies facilitated imperial expansion, power consolidation, and economic exploitation by marginalizing IKSs (Ciófalo & Ortiz-Torres, 2024). Moreover, to sustain this exploitation, colonial powers enforced rigid sexual hierarchies, creating power imbalances that dehumanized vast segments of the global population by reducing them to mere (re)productive labor for the benefit of imperial interests (Nassiri-Ansari & Rhule, 2024). Indigenous women's roles were redefined to fit colonial expectations, often relegating them to subordinate positions and framing them as inferior beings within both colonial and Indigenous societies (Lieu et al., 2020; Nassiri-Ansari & Rhule, 2024). Colonial racial and gender constructs intersected to create enduring power dynamics and inequities (Nassiri-Ansari & Rhule, 2024). Moreover, colonial powers disrupted Indigenous women's economic autonomy by introducing capitalist production models, shifting resource control away from Indigenous communities, particularly women, and marginalizing their economic roles. Such entrepreneurship can thus be rethought as a necroentrepreneurial practice, underpinned by Necropolitics (Mbembe, 2003). Mbembe's necropolitics describes how oppressive systems control populations through violence, torture, slavery, and warfare, treating colonized peoples as disposable or without intrinsic value (Imas & Garcia-Lorenzo, 2023). Within this framework, Africans had no value except for the benefit and exploitation of white colonialists (Imas & Garcia-Lorenzo, 2023). Today, the colonial legacy persists through patriarchal and capitalist systems (Lokot et al., 2024), which prioritize large-scale, market-driven economies that undermine IE.

Therefore, decolonial feminism is employed to bring to bear Indigenous women's perspectives in entrepreneurship. This theory challenges mainstream entrepreneurship's colonial and patriarchal assumptions (Imas & Garcia-Lorenzo, 2023). That is, entrepreneurship is shown not as a neutral or universal phenomenon, but as embedded within and perpetuating patriarchal, capitalist, and heteronormative systems. Moreover, the theory challenges the dominant narrative of IE as neutral spaces, revealing them instead as arenas deeply influenced by neoliberal capitalist ideologies. For instance, despite post-1994 reforms in South Africa and broader post-colonial efforts in Africa, entrepreneurship remains limited by masculine norms and neoliberal capitalist logics. Decolonial feminism offers a critical lens through which to understand entrepreneurship by grounding it within the cultural, ideological, and historical contexts of African women's lives. Moreover, decolonial feminism redefines Indigenous women's entrepreneurship as active resistance against oppressive structures, enabling an exploration of their motivations and unique challenges. Ultimately, this study enriches scholarship advocating for epistemic justice, Indigenous self-determination, and feminist economic alternatives in the Global South.

On Rethinking Indigenous Entrepreneurship

Entrepreneurship remains rooted in Eurocentric narratives that mask neo-colonial discourse and practice behind concepts such as business creation, organization, and innovation (Imas & Garcia-Lorenzo, 2023). This entrepreneurial narrative hides the colonial past upon which European enterprises were built through slavery and oppression of individuals from the global South and therefore sustains socioeconomic oppression (Imas & Garcia-Lorenzo, 2023). Particularly among Indigenous communities who constitute approximately 5–6% (~476 million) of the global population, spanning 90 countries, half of whom are women (Padilla-Meléndez et al., 2022). Entrepreneurship has been touted in Africa as a solution to poverty and other societal ills (Sall, 2022). However, such a characterization preserves a neo-colonial agenda, and this, in turn, leads to the disempowerment of IKSs and a silencing of other narratives, particularly women's narratives, through the imposition of external entrepreneurial frameworks that pay no attention to contextual specifics and lived experiences. As Imas and Garcia-Lorenzo (2023) argue, these colonial and imperial ideologies must be decolonized to liberate other forms of entrepreneurship that are already being carried out by subaltern communities in the Global South. Beresford (2020) also condemns this enforced celebration of entrepreneurship as a deadweight, particularly in countries like South Africa, where communities have alternative economic imaginations. Hence, this study argues for the acknowledgement and advancement of IE models as a means of reclaiming economic agency and cultural sovereignty.

IE is conceptualized as the establishment, management, and development of new business ventures by Indigenous people based on IKS to serve their own communities (Padilla-Meléndez et al., 2022), in this case Nguni and Abathwa, and where the benefits may extend to non-indigenous networks at large. Therefore, Indigenous entrepreneurs are people who build and operate such entities (Padilla-Meléndez et al., 2022). Indigenous entrepreneurs are seen as catalysts for rural development, driving economic growth and fostering community cohesion by leveraging local resources (Cheteni & Umejesi, 2024). IE emphasizes the utilization of IK, skills, and expertise to preserve cultural integrity and for local economic development (Cheteni & Umejesi, 2024). In contrast to mainstream entrepreneurship, which is typically aimed at individual or corporate profit. IE prioritizes community empowerment, cultural preservation, environmental protection, and sustainable development (Cheteni & Umejesi, 2023; Padilla-Meléndez et al., 2022; Tshikovhi, 2021). That is, profitability remains a factor, but Indigenous entrepreneurs juxtapose it with social and environmental considerations.

Indigenous entrepreneurs possess a special advantage in accessing the epistemological heritage of their people and, in the process, preserving it for future generations. That is, IE requires an explicit engagement with indigeneity itself, not as a cultural artefact, but as a dynamic identity that shapes lived experience, access to power, and knowledge production (Cheteni & Umejesi, 2024). Indigenous entrepreneurs' strong connection to land, secured through ancestral ties, often leads them to adopt environmentally sustainable practices. This is reinforced by economic incentives when the environment is an integral part of the product or service they offer (Cheteni & Umejesi, 2023). That is, for Indigenous communities, the recovery and utilization of ancestral land and related resources serve not just as cultural foundations but also as essential resources for entrepreneurial ventures (Tshikovhi, 2021). In the South African context, perhaps Indigenous entrepreneurs are driven by the spirit of '*ubuntu*' to consider both natural and human aspects in their businesses (Selelo, 2023).

Indigenous business practices have existed for centuries (Demong et al., 2023). For instance, Indigenous communities in Latin America and Africa have long been involved in barter trading (Cheteni & Umejesi, 2024). Far from being static or outdated, Indigenous entrepreneurs are dynamic and create economically viable solutions that are useful to all populations and solve contemporary issues. A notable case is the production of shea butter, which has been traditionally used in skincare and medicine in sub-Saharan Africa, and has since gained universal recognition across various industries, such as cosmetics, pharmaceuticals, and chocolate formulations (Onwuegbuzie & Mafimisebi, 2021). In Indigenous contexts, entrepreneurship is not just an economic activity but also a path to achieving self-determination and sovereignty (Gouvea et al., 2022). By promoting locally driven innovation and reducing state dependency, IE builds community agency and promotes more inclusive and sustainable economic systems (Macpherson et al., 2021). Therefore, this chapter argues that IE empowers marginalized communities and builds resilient entrepreneurial ecosystems. Although entrepreneurship is widely recognized as a catalyst for development in Sub-Saharan Africa, Indigenous entrepreneurs' specific contributions and practices remain underexplored (Eijdenberg et al., 2024). This is despite the prevalence of diverse Indigenous entrepreneurial activities across the region. Gaining a deeper understanding of IE can offer valuable insights for both policymakers and business practitioners, potentially informing more culturally attuned and sustainable business models (Cheteni & Umejesi, 2023).

Indigenous Women Entrepreneurship and Empowerment

The concept of 'minority entrepreneurship' encapsulates the engagement of individuals from non-dominant groups, including class, race, age, gender, sexual orientation, or ethnicity, in business activities (Vorobeva, 2022). Within this framework, IWE emerges as a distinct area of study, intertwining aspects of indigeneity, gender, and entrepreneurial practice (Croce, 2020). Entrepreneurship is traditionally considered to be primarily a male activity, marginalizing the contributions and potential of Indigenous women entrepreneurs (Croce, 2020; Vorobeva, 2022). Yet, Indigenous women often demonstrate a higher propensity for initiating entrepreneurial ventures compared to their male counterparts (Demong et al., 2023). Furthermore,

Indigenous women participated in entrepreneurial activities during the colonial period, despite limited access to resources, including land (Shash, 2021). While colonial powers concentrated on expanding territorial dominance, Indigenous women were actively exercising economic agency by engaging in trade and operating within informal markets (Shash, 2021). Through entrepreneurship, these women achieved economic independence and redefined their roles in patriarchal family structures, thus challenging traditionally held power relations with elders and husbands (Shash, 2021). That is, women navigate multiple oppressive systems while refusing to be fully defined by them. This demonstrates that entrepreneurship can be a vehicle not only for economic empowerment but also for reconfiguring gender hierarchies rooted in colonial and customary systems.

IWE, therefore, is not just an academic topic; it represents a movement towards recognizing Indigenous women's entrepreneurial activities as avenues for self-emancipation, leadership development, and community well-being through economic and social empowerment (Croce, 2020). This perspective aligns with global efforts to achieve gender equality and eliminate gender-based violence, acknowledging Indigenous women's pivotal role in attaining socioeconomic development and leading emancipation processes within their communities (Croce, 2020). Indigenous women in Zambia, Gambia, Kenya, Nigeria, Tanzania, Ghana, and South Africa are significantly represented in the small and medium enterprise sector (Gwandure & Lukhele-Olorunju, 2023). Despite their active involvement in small business development, scholarly literature on their experiences and contributions remains limited (Gwandure & Lukhele-Olorunju, 2023). In South Africa, for instance, rural Indigenous women have significantly contributed to the nation's economy through entrepreneurial ventures that provide income, employment, and support for local cultural resources (Donga & Chimucheka, 2024). Yet, their impact remains under-researched, highlighting the need for more inclusive and context-sensitive studies that center their voices and acknowledge their contributions to local development. This study is a starting point to further discuss issues across the broader gender spectrum, and thus, it contributes to the missing literature on gender perspectives within IE. There is a need to recognize and advance Indigenous women's entrepreneurship to create inclusive socioeconomic development and empower marginalized communities.

Indigenous Women's Entrepreneurial Activities in KwaZulu-Natal

The findings reveal that Indigenous women entrepreneurs in rural KZN actively engage in diverse agricultural ventures, including crop cultivation and livestock rearing. One of the respondents emphasized agriculture's essential role in community survival, stating, "Agriculture is not fashion. Everyone must eat; food is always needed" (respondent 2). This sentiment echoes Groenewald's (1993) argument that the sustainability of African agriculture is increasingly dependent on entrepreneurial innovation. Another respondent shared that she preserves and sells leafy greens like *imifino* (wild spinach) through a traditional method known as "*ifantane*," ensuring food security through seasonal scarcity and sustained household income (respondent 7). In addition to crop farming, several women operate poultry businesses to meet nutritional needs and support household income. Another respondent extends this model by breeding and selling pigs and supplying specific chicken breeds to traditional healers for use in rituals (respondent 6). For example, *Impangela* chickens (Potchefstroom Koekoek) are used in ancestral rites, while *Ezimunyama* chickens (Black Australorp) are important for purification ceremonies (Tenza et al., 2024). Across Africa, it is women who lead chicken production, as this business offers them essential nutritional, economic, and cultural benefits (Tenza et al., 2024).

Some Indigenous women entrepreneurs engage in culturally based handicrafts. One, for instance, weaves grass mats (respondent 4), and another produces beaded jewellery for women that is used mainly in major rites of passage like *umemulo* (coming-of-age ceremony) or *umshado* (marriage) (respondents 5 & 7). Beadwork, traditionally associated with identity and spirituality, has become a source of livelihood and the preservation of cultural heritage, thus opening opportunities for self-employment and artisan empowerment among rural women. Women's dominance in handicrafts was expected, aligning with the United Nations data indicating that 56.3% of artisans globally are women (Parthiban et al., 2024). Within the framework of traditional healing, one of the respondents is *umelaphi wendabuko* (traditional healer), practising various healing approaches using *umuthi* (medicinal plants), prayer, and water (respondent 3). Another respondent practises traditional midwifery and prepares *isihlabezo*, a pregnant woman's herbal tonic (respondent 9). This is not surprising as traditional healing remains relevant in marginalized communities, where mainstream healthcare services are scarce. In fact, more than 65% of poor African Indigenous people depend on IKS for their health issues (Onwuegbuzie & Mafimisebi, 2021). This fact is corroborated by the World Health Organization, which reports that 80% of South Africans use traditional medicine for their overall health requirements (Tshamano et al., 2023).

It is evident that Indigenous entrepreneurs focus on "at hand resources and strategies", and this makes IE distinct (Onwuegbuzie & Mafimisebi, 2021; Selelo, 2023). These entrepreneurial ventures demonstrate women's capacity to sustain their livelihoods and that this capacity is deeply intertwined with the preservation, application, and integration of IKS in entrepreneurial ventures. IKS is indeed a critical tool for entrepreneurship, and Indigenous culture generates a series of goods and services for which women have a comparative advantage or an entrepreneurial market niche. For instance, one of the respondents shared, "I am the only person in this area who weaves grass mats; that is why they come to me, they cannot go to anyone else" (respondent 4). While some women pursue individual entrepreneurial activities, others, such as those involved in poultry farming, exemplify collective action by pooling resources and capabilities to generate shared socioeconomic value. That is, poultry cooperatives exemplify a collectivist culture of indigenous people (Tenza et al., 2024). The findings reaffirm that women work together and actively co-create sustainable livelihoods rooted in communal values such as reciprocity, cooperation, and mutual support (Bansal et al., 2024), which the colonial capitalist tried to erase. That is, they are reclaiming collective stewardship over community resources and fostering the resurgence of a decolonized Indigenous identity.

In doing so, these women also disrupt dominant stereotypes that depict Indigenous entrepreneurs as traditional, insular, and reactionary (Casey *et al.*, 2024). Onwuegbuzie and Mafimisebi (2021) argue that while their innovations may appear rudimentary, they are functionally robust and contextually appropriate, showcasing the adaptive resilience of Indigenous women in resource-constrained settings. The labelling of their ventures as rudimentary reinforces the coloniality of power that renders Indigenous women invisible in dominant economic paradigms. This invisibility is precisely where decolonial possibility lies, because it opens space for alternative, relational, and collective modes of doing economy. Moreover, IE is devalued and invisible in the formal economy, and circumstances are amplified for women due to their double status as women and entrepreneurs. Their enterprises are categorized as informal, and women appear restricted to informality due to a lack of access to formal business opportunities and the continuing influence of traditional gender roles, which restrict women to household responsibilities (Demong et al., 2023). This is also the case in Malaysia, where similar gender-specific expectations restrict women to domestic activities such as agricultural and small-scale manufacturing, thus hindering their access to entrepreneurial resources and opportunities (Demong et al., 2023).

Most Indigenous women entrepreneurs finance their ventures through personal savings. The women also demonstrate pluractivity. Pluriactivity is a term used to describe an individual who engages in various income-generating activities (Mthembu, 2012) to maintain and/or expand their principal enterprises. For example, one of the respondents sells Kudu tea imported from China so that she can reinvest the capital into her beadwork (respondent 7). Only a few people get seed financing from government institutions alongside business training and advisory support. For instance, the women involved in poultry farming were trained and received seed funding, as did those engaged in beadwork. These examples highlight the capacity and potential of Indigenous women entrepreneurs when afforded institutional support, directly countering dominant narratives that question their capabilities.

For instance, Western institutions like the International Labour Organization claim African women are unable or in no position to head business ventures (Imas & Garcia-Lorenzo, 2023), echoing the colonial denial of African women's agency. These assumptions are reductionist and oblivious to the significant roles played by women in African economies, as illustrated by the Nguni and Abathwa Indigenous communities. In fact, in precolonial times, women in non-Western cultures often held powerful positions that did not conform to strictly masculine or feminine roles (Amadiume, 1987; Chineme et al., 2022). Many African societies were shaped by strong matriarchal traditions that granted women considerable influence and authority (Chineme et al., 2022). In countries such as Nigeria (Zazzau, Nupe, and Edo), the Republic of Benin (Dahomey), Sierra Leone (Mende), Ghana (Ashanti), Ethiopia (Sheba), Sudan (Kandake), and Angola (Nzinga), women's authority was integral to governance and community life (Chineme et al., 2022). Matriarchal and patrilineal systems often coexisted without diminishing the status of women, and Indigenous African women held prominent titles and leadership roles, including queen-rulers, queen mothers, paramount chiefs, Iyalodes, Sagi, and Sonya, reflecting a long-standing recognition of their political and cultural significance (Chineme et al., 2022). That is, gendered activities and responsibilities were understood as more fluid and flexible than in Western society. In support of this, Moagi and Mtombeni (2021) assert that southern African women were not confined to domestic roles in precolonial times. Instead, they held active economic, political, and religious roles. They participated in domains often associated with men and exercised autonomy over their own bodies and labor. Indigenous women entrepreneurs, therefore, deconstruct the myth about a heroic, white, young, male entrepreneur with their own examples of entrepreneurial ventures. Through the lens of decolonial feminism, it is clear that their ventures are not merely economic activities but symbolic acts of resistance to the colonial/modern gender system. Considering this, IWE should be understood as a form of decolonial feminist practice that reclaims women's agency, revalues ways of knowing often dismissed by Eurocentrism, and pushes back against the intertwined legacies of patriarchy, colonialism, and capitalism. These women are not simply earning a living; they are affirming IKSs, challenging gender norms, and offering new, culturally grounded visions of what entrepreneurship can look like in African societies.

Motivational Drivers of Indigenous Women Entrepreneurs

The sustainability and success of Indigenous businesses are linked to entrepreneurial motivation, which is the driving force that shapes an entrepreneur's ability to start, maintain, and expand a business (Fulgence, 2024). This motivation is often driven by a range of factors, including a desire to create value, address problems, enjoy autonomy, and embrace risks, among other factors (Fulgence, 2024). Studies have found that women entrepreneurs are motivated by different factors to get into entrepreneurship compared to their male counterparts (Demong et al., 2023). Men are mainly motivated by the prospects of profits and financial status (Demong et al., 2023). While women are often involved in entrepreneurship out of necessity (Okeke-Uzodike, 2019), as is the case for some respondents in this study. For instance, one respondent shared, "we are not only about profit; it is also about taking care of our households" (respondent 6). Another respondent emphasized the significance of intrinsic motivation when she shared, "some people fail in this business because they lack love for this work." Without passion, it is easy to give up" (respondent 1). Through the decolonial feminism lens, these views contradict mainstream economic discourse, emphasizing profit as the key indicator of success. Instead, they disclose a gendered ethic of care, survival, and affective labour as definitive of Indigenous women's entrepreneurial agency. That is, women's capacity for entrepreneurship is based on a desire to survive, reflecting a creative, precarious resistance marked by ongoing negotiation against marginalization and an insistence against being subjected to dependency. This rejection of dependency reflects active disobedience to colonial gender scripts. Moreover, women show high levels of self-reliance, driven by their roles as caregivers and their need to provide for their families where reliable male economic support is unavailable. Furthermore, studies indicate that women tend to channel business profits back into areas such as education, healthcare, and community upliftment, thereby increasing the broader social impact of their entrepreneurial activities (Ebewo et al., 2025).

This ethic of necessity and resilience is further shaped by broader structural conditions. While some material conditions have improved since the colonial era, many Indigenous communities remain economically and geographically marginalized, with limited access to formal employment and continued dependence on local resources (AlMehrzi et al., 2024;

Colbourne, 2017). These persistent legacies of colonialism, including poverty, exclusion, and exploitation, often compel women into informal entrepreneurial activities as a means of survival and self-determination. As such, they can be characterized as 'underdog' entrepreneurs, individuals who face multiple intersecting barriers, including socio-cultural, economic, cognitive, and physical challenges (Menzies et al., 2024). This, in turn, cultivates distinct entrepreneurial traits such as resilience, creativity, discipline, and robust social networks (Menzies et al., 2024), which underpin the adaptive capacities of Indigenous women in navigating constrained environments.

IE is often driven by the long-term vision of preserving cultural heritage and transmitting values (Fulgence, 2024) to sustain a specific way of life despite experiencing the effects of cultural annihilation due to colonization. The entrepreneurs embed cultural relevance in their products and services, creating experiences for customers seeking authenticity and connection to Indigenous worldviews (Cheteni & Umejesi, 2024). In doing so, this strategy creates connections beyond typical market exchanges by prioritizing cultural continuity, collective well-being, and environmental sustainability (Cheteni & Umejesi, 2024). Therefore, Indigenous communities still employ entrepreneurship as a political project, a force for economic recovery and cultural revitalization. In this case, culture is both a driver and a strategic asset. That is, identity and tradition inform both vision and craft in Indigenous enterprises (Fulgence, 2024).

The woman involved in the poultry business explained that in times of financial hardship, such as during a funeral, she allows community members to take chickens on credit, with payment deferred (respondent 1). This practice exemplifies the *ubuntu* ethic, where entrepreneurship is not solely profit-driven but embedded in communal care and reciprocity. Such acts highlight how Indigenous women's enterprises are vehicles for both economic participation and social solidarity. Moreover, the participation of women in entrepreneurship works to improve their economic conditions while also challenging and redefining deeply entrenched gender norms. In Algeria, the involvement of women in business has prompted dramatic socio-cultural changes within the broader context of the empowerment movement (Shash, 202). Thus, entrepreneurship becomes a political act that challenges patriarchal assumptions regarding men's central roles as persons of economic and cultural power (Imas & Garcia-Lorenzo, 2023). The women employ spiritual, ecological, and economic logics, thereby asserting their autonomy in their economies and communities, respectively. These actions contest the Eurocentric capitalist approach to entrepreneurship as merely making profits and formalization.

Challenges Facing Indigenous Women Involved in Entrepreneurship

The challenges confronting Indigenous women entrepreneurs in rural South Africa are deeply entrenched in the intertwined legacies of colonialism, capitalism, and patriarchy, which Lugones (2010) describes as the 'coloniality of gender.' Respondents highlighted a range of systemic challenges that severely limit their entrepreneurial potential. These challenges include a persistent lack of institutional support, as the business landscape in South Africa remains predominantly male-dominated (Diale & Carrim, 2022). Women entrepreneurs continue to face systemic disadvantages, including limited access to capital, economic networks, and social recognition (Donga & Chimucheka, 2024). The lack of consistent, accessible support in South Africa has led to frustration and diminished confidence among Indigenous entrepreneurs (Selelo, 2023). Additional barriers include the high cost of transportation, which significantly reduces profit margins (respondent 1), and restricted access to land, often controlled by traditional leaders whose decisions are influenced by localized power dynamics and intra-community politics (respondent 2). Inadequate infrastructure, particularly insufficient water supply, further hinders agricultural productivity (respondent 2). Moreover, the lack of formal market infrastructure forces many women to rely on informal transactions, exposing them to financial vulnerability when customers fail to honor payments for goods like beadwork and jewellery (respondent 5). The remoteness of Indigenous communities exacerbates these constraints, limiting access to markets, infrastructure, and social networks (Selelo, 2023). One respondent shared, "I am self-reliant. My main difficulty is that I cannot access the river to harvest *ncema* grass, so I have to buy it instead" (respondent 4), illustrating how resource inaccessibility and environmental constraints compound financial burdens. These lived realities reveal how Indigenous women's entrepreneurial agency is persistently undermined by systemic marginalization, infrastructural neglect, and enduring gendered inequalities embedded in the postcolonial state.

A decolonial feminist analysis reveals that IWE is not simply hindered by economic underdevelopment. On the one hand, Indigenous women are forced to accept roles and identities socially engineered under colonization, which strengthens patriarchal rule (Imas & Garcia-Lorenzo, 2023) and subordinates them by design. The fact that women are still dependent on patriarchal gatekeepers such as traditional leaders for land access illustrates how neocolonial patriarchal rule persists under the guise of "custom," often protected even within postcolonial states. That is, power is hoarded and gendered within traditional leadership systems, legacies of indirect colonial rule now entrenched in postcolonial governance structures. Moreover, the marginalization of Indigenous women entrepreneurs within the broader economic system reflects the racialized and gendered contours of capitalism. The dominant image of the entrepreneur, as white, male, and middle-class, renders Indigenous women invisible or illegitimate in the market. They are not viewed as equal actors by other entrepreneurs, customers, and institutions, which leads to exclusion. This exclusion manifests in poor access to capital, training, infrastructure, and networks. It also means that marketing, a vital function for business growth (Selelo, 2023), is structurally inaccessible or underdeveloped. Poor marketing capabilities then translate into underperformance, reinforcing economic stagnation and invisibility.

A decolonial feminist lens foregrounds how these women's struggles are shaped by intersecting oppressions and how their economic participation is simultaneously a demand for cultural recognition, gender equity, and structural transformation. Moreover, these are not isolated failures but features of a system designed to maintain hierarchies of power, wherein Indigenous women's work is simultaneously devalued, exploited, and excluded. The triple marginalization of these women, as Indigenous, as women, and as informal actors, reflects "necroentrepreneurship," where entrepreneurial activity is allowed only insofar as it remains precarious, under-resourced, and disconnected from formal structures of support. Importantly, decolonial feminism also sees entrepreneurship as resistance. These women's microstorias (personal, lived stories of resistance) highlight forms of survival, creativity, and self-determination that contest both colonial-capitalist norms and internalized patriarchy. When a respondent says, "I am self-reliant… I buy *ncema* grass," this is more than economic adaptation; it is a statement of autonomous agency within a system that persistently undermines it. Furthermore, cultural expectations require women to take on unpaid caregiving work within domestic and community environments, and this continues to tether Indigenous women to the domestic sphere, obstructing their entrepreneurial potential. As Lugones (2010) asserts, patriarchy under coloniality racialized and feminized labor, reducing women to social roles that facilitate capital accumulation while denying them recognition and resources. These gendered expectations create a self-sustaining cycle of marginalization that maintains Indigenous women in poverty, excluding them from viable economic participation (Quagrainie & Kabalan, 2024). It is also clear that cultural values and traditions can facilitate and constrain entrepreneurial activities (Menzies et al., 2024; Padilla-Meléndez et al., 2022).

Another challenge noted is that older women dominate IE. Most young Indigenous people lack the resources, entrepreneurial capabilities, and institutional support needed to translate their business concepts into successful enterprises (Lose & Cheteni, 2024). A study by Menzies et al. (2024) supports these study findings and identifies some of the systemic and structural challenges facing Indigenous women entrepreneurs to be racial and gender discrimination, limited government and institutional support, limited business and management education, technological exclusion, assimilation to the mainstream economy, and erosion of self-esteem and motivation. In another study, Demong et al. (2023) discuss how historical and current colonial marginalization, limited financial capital, and cultural and linguistic disconnections remain significant impediments to IE development. Indigenous women in other instances experience gender-based violence, sexual and physical violence, and psychological violence (Croce, 2020).

Conclusion

To conclude, Africa and South Africa in particular grapple with the twin imperatives of inclusion and sustainability; therefore, development efforts must shift towards models that reflect marginalized communities' lived experiences and aspirations (Tshamano et al., 2023). In this context, Indigenous women in KZN, though persistently marginalized by state structures, capitalist markets, and internal community hierarchies, are not passive recipients of development but active agents of transformation. They actively engage in diverse ventures and therefore redefine entrepreneurship in culturally grounded, collectivist, and care-centered terms, offering alternative visions of economic participation rooted in Indigenous worldviews. Their motivations reflect a gendered ethic of care, where entrepreneurship is not simply about income, but about sustaining households, communities, traditions, and the environment. Drawing from IKS, they engage in forms of entrepreneurship that prioritize community well-being, environmental stewardship, and cultural continuity, values largely dismissed in neoliberal and Eurocentric entrepreneurship models that privilege profit maximization. Therefore, integrating IKS into small enterprise development is vital for crafting inclusive and ecologically sustainable rural development (Cheteni & Umejesi, 2024). Their ventures, from poultry farming and beadwork to traditional healing and land-based microbusinesses, contest the dominant narrative of entrepreneurship as inherently formal, masculine, and profit-driven.

Through the decolonial feminism lens, IWE emerges not as a marginal or supplemental economic pursuit but as a deeply political practice shaped by colonial legacies, patriarchal domination, capitalist exclusion, and epistemic violence. In this context, the notion of an 'Indigenous model' of entrepreneurship represents a rejection of necroentrepreneurial systems that commodify and devalue Indigenous lives (Imas & Garcia-Lorenzo,

2023). These ventures contribute to economic decolonization by affirming Indigenous rights, sovereignty, and self-determination. In doing so, they create alternative economic systems that prioritize social, cultural, environmental, and economic value from an Indigenous-centric perspective (Colbourne et al., 2024). That is, women are not passive victims of underdevelopment, and entrepreneurial activities are far more than economic survival strategies; they are acts of cultural promotion/resistance, identity reclamation, and gendered defiance against layered systems of oppression. Women have reimagined entrepreneurship through diverse, pluralistic lenses. Moreover, their economic participation serves as a vehicle for unsettling entrenched patriarchal norms within households and communities. Entrepreneurship becomes a site of gender renegotiation, a strategic means through which Indigenous women assert agency, reclaim authority, and challenge colonial gender scripts that historically confined them to subordinate roles.

Despite the lack of institutional support, the prevalence of unpaid care burdens, and the dominance of neoliberal policies that erode local economies, Indigenous women demonstrate the possibilities of entrepreneurship as a tool for decolonization, economic empowerment, and feminist transformation. Their practices call for urgent rethinking of entrepreneurship policy in South Africa, one that recognizes and supports the unique knowledge systems, needs, and values of Indigenous women. That is, entrepreneurship can serve as a powerful tool for livelihoods without being exploitative or extractive. To achieve this, entrepreneurial initiatives must adopt a decolonial approach, centered on relational values such as reciprocity, collective well-being, and environmental sustainability. This requires moving beyond profit-maximization models rooted in capitalist logics and instead grounding enterprise development in Indigenous epistemologies. When development strategies are shaped by the lived knowledge, cultural priorities, and socio-environmental ethics of Indigenous communities, they foster inclusive growth while resisting resource plunder and social marginalization.

To move toward gender equity and sustainable socioeconomic development, policy interventions must go beyond token inclusion and instead work to dismantle the structural, historical, and epistemological barriers that continue to marginalize Indigenous women's knowledge, labor, and leadership. However, achieving meaningful economic empowerment for Indigenous women demands holistic strategies that confront not only structural inequalities but also deep-rooted cultural and educational exclusions. Therefore, policy responses must move beyond one-size-fits-all frameworks and instead prioritize targeted interventions for those most marginalized. This requires recognizing IE not as peripheral but as central to any truly inclusive, just, and sustainable economic future. Globally, targeted financial initiatives are reshaping IE. In Latin America, the "Fondo de Apoyo a los Pueblos Indígenas (IPAF)" offers two-year grants of US$20,000–US$50,000 to support Indigenous economic development (Gouvea et al., 2022). In Brazil, this funding helped the Kayaby community preserve and commercialize native seeds. Such culturally responsive funding models offer valuable lessons for Africa and regions like KZN, where tailored support could empower Indigenous women and foster inclusive development. Ultimately, advancing IWE requires a coordinated, multisectoral approach. Government agencies, academic institutions, non-governmental actors, and private actors must collaborate to create enabling environments.

References

Adler, E. (1997). Seizing the middle ground: Constructivism in world politics. *European Journal of International Relations, 3*(3), 319–363. https://doi.org/10.1177/1354066197003003003

AlMehrzi, A. A., Tipu, S. A., & Sarker, A. E. (2024). Determinants, processes, and impacts of indigenous entrepreneurship: A systematic literature review. *Journal of Enterprising Communities: People and Places in the Global Economy, 18*(3), 646–681. https://doi.org/10.1108/JEC-10-2022-0155

Amadiume, I. (1997). *Re-inventing Africa: Matriarchy, religion and culture.* Zed Books.

Bansal, S., Sarker, T., Yadav, A., Garg, I., Gupta, M., & Sarvaiya, H. (2024). Indigenous communities and sustainable development: A review and research agenda. *Global Business and Organisational Excellence, 43*(4), 65–87. https://doi.org/10.1002/joe.22237

Beresford, M. (2020). Entrepreneurship as legacy building: Reimagining the economy in post-apartheid South Africa. *Economic Anthropology, 7*(1), 65–79. https://doi.org/10.1002/sea2.12170

Braun, V., & Clarke, V. (2006). Using thematic analysis in psychology. *Qualitative Research in Psychology, 3*(2), 77–101. https://doi.org/10.1191/1478088706qp063oa

Casey, S., Crimmins, G., McIntyre, J., & O'Sullivan, S. (2024). Stories of country women: Advancing feminist intersectionality for climate change adaptation in Australia. *Sociologia Ruralis, 29*(3), 1–15. https://doi.org/10.1111/soru.12474

Cele, N. (2023). *A decolonial feminist investigation of gender representation in isiZulu literature in the Further Education and Training Phase* (Master's thesis, University of KwaZulu-Natal).

Cheteni, P., & Umejesi, I. (2023). Application of indigenous knowledge and sustainable agriculture by small-scale farmer households and entrepreneurs in South Africa. *International Journal of Development and Sustainability, 12*(1), 1–18.

Cheteni, P., & Umejesi, I. (2024). Indigenous knowledge in entrepreneurship and cultural tourism in the rural areas. *Comparative Sociology, 23*(3), 340–359. https://doi.org/10.1163/15691330-bja10110

Chineme, A., Assefa, G., Herremans, I. M., Wylant, B., & Shumo, M. (2022). African Indigenous female entrepreneurs (IFÉs): A closed-looped social circular economy waste management model. *Sustainability, 14*(18), Article 11628. https://doi.org/10.3390/su141811628

Ciófalo, N., & Ortiz-Torres, B. (2024). Toward decolonial community psychologies from Abya Yala. *American Journal of Community Psychology*, 1–12. https://doi.org/10.1002/ajcp.12746

Colbourne, R. (2017). Indigenous entrepreneurship and hybrid ventures. In A. Corbett & J. Katz (Eds.), *Perspectives and approaches to blended value entrepreneurship: Advances in entrepreneurship, firm emergence and growth* (Vol. 19). Emerald Publishing.

Colbourne, R., Ejaz, L., Grinevich, V., Husain, S., & O'Farrell, D. A. (2024). Hybrid entrepreneurship in just transitions: Dealing with dilemmas facing 'the other'. *Environmental Innovation and Societal Transitions, 53*, Article 100924. https://doi.org/10.1016/j.eist.2024.100924

Creswell, J. W., & Creswell, J. D. (2005). *Mixed methods research: Developments, debates, and dilemmas.* Berrett-Koehler Publishers.

Croce, F. (2020). Indigenous women entrepreneurship: Analysis of a promising research theme at the intersection of Indigenous and women entrepreneurship. *Ethnic and Racial Studies, 43*(6), 1013–1031. https://doi.org/10.1080/01419870.2019.1630659

Demong, N. A. R., Kassim, E. S., Ibrahim, I., Zamzuri, N. H., Omar, E. N., & Jalil, S. A. (2023). Understanding Indigenous entrepreneurship in Malaysia: A gender perspective. *Journal of Emerging Economies and Islamic Research, 11*(2), 75–93.

Diale, D., & Carrim, N. M. (2022). Experiences of Black African women entrepreneurs in the South African male-dominated entrepreneurial environments. *Journal of Psychology in Africa, 32*(3), 216–223. https://doi.org/10.1080/14330237.2021.2018109

Donga, G., & Chimucheka, T. (2024). Rural women entrepreneurship in South Africa: Issues and challenges. *African Journal of Inter/Multidisciplinary Studies, 6*(1), 1–12. https://doi.org/10.51415/ajims.v6i1.1175

Ebewo, P. E., Schultz, C., & Mmako, M. M. (2025). Towards inclusive entrepreneurship: Addressing constraining and contributing factors for women entrepreneurs in South Africa. *Administrative Sciences, 15*(1), 1–20.

Eijdenberg, E. L., Ostertag, F., Dana, L. P., & Onwuegbuzie, H. (2024). The significance of Indigenous entrepreneurship in Sub-Saharan Africa: An introduction. In E. L. Eijdenberg, F. Ostertag, L. P. Dana, & H. Onwuegbuzie (Eds.), *Indigenous entrepreneurship in Sub-Saharan Africa: Implications for theory and practice.* Palgrave Macmillan.

Fulgence, K. (2024). Entrepreneurial motivation among small-scale business entrepreneurs: A case study of Tanzania. In E. L. Eijdenberg, F. Ostertag, L. P. Dana, & H. Onwuegbuzie (Eds.), *Indigenous entrepreneurship in Sub-Saharan Africa: Implications for theory and practice.* Palgrave Macmillan.

Gouvea, R., Lehneman, L., & Terra, B. (2022). Tribal economic development and entrepreneurship: A Latin American perspective. *World Development Perspectives, 26*, Article 100403. https://doi.org/10.1016/j.wdp.2022.100403

Groenewald, J. A. (1993). Resource use entitlement, finance and entrepreneurship in Sub-Saharan smallholder agriculture. *Agrekon, 32*(3), 122–128.

Gwandure, C., & Lukhele-Olorunju, P. (2023). Women's use of Indigenous knowledge in Africa. *Social Sciences & Humanities Open, 8*(1), Article 100741. https://doi.org/10.1016/j.ssaho.2023.100741

Imas, J. M., & Garcia-Lorenzo, L. (2023). A postcolonial and Pan-African feminist reading of Zimbabwean women entrepreneurs. *Gender, Work & Organization, 30*(2), 391–411. https://doi.org/10.1111/gwao.12931

Lieu, J., Sorman, A. H., Johnson, O. W., Virla, L. D., & Resurrección, B. P. (2020). Three sides to every story: Gender perspectives in energy transition pathways in Canada, Kenya and Spain. *Energy Research & Social Science, 68*, Article 101550. https://doi.org/10.1016/j.erss.2020.101550

Lokot, M., Pichon, M., Kalichman, B., Nardella, S., Falconer, J., Kyegombe, N., & Buller, A. M. (2024). Decolonising the field of violence against women and girls: A scoping review and recommendations for research and programming. *Social Science & Medicine*, Article 117168. https://doi.org/10.1016/j.socscimed.2024.117168

Lose, T., & Cheteni, P. (2024). Readiness for new ventures in South Africa through entrepreneurship education: A reflective thematic approach. *Cogent Business & Management, 11*(1), Article 2315693. https://doi.org/10.1080/23311975.2024.2315693

Lugones, M. (2010). Toward a decolonial feminism. *Hypatia, 25*(4), 742–759.

Macpherson, W. G., Tretiakov, A., Mika, J. P., & Felzensztein, C. (2021). Indigenous entrepreneurship: Insights from Chile and New Zealand. *Journal of Business Research,* 127, 77–84. https://doi.org/10.1016/j.jbusres.2021.01.013

Majokweni, P., Petersen, I. H., Mayeki, S., & Maziya, M. (2024). *Gender, informality and innovation: Insights from research in Sweetwaters, KwaZulu-Natal* [Research report].

Mbembe, A. (2003). Necropolitics. *Public Culture, 15*(1), 11–40. https://doi.org/10.1215/08992363-15-1-11

Menzies, J., Chavan, M., Jack, R., Scarparo, S., & Chirico, F. (2024). Australian Indigenous female entrepreneurs: The role of adversity quotient. *Journal of Business Research, 175*, Article 114558. https://doi.org/10.1016/j.jbusres.2024.114558

Mignolo, W. D. (2007). Introduction: Coloniality of power and decolonial thinking. *Cultural Studies, 21*(2–3), 155–167. https://doi.org/10.1080/09502380601162498

Moagi, L. A., & Mtombeni, B. (2021). In O. Yacob-Haliso & T. Falola (Eds.), *The Palgrave handbook of African women's studies*. Palgrave Macmillan.

Mthembu, M. B. J. (2012). Rural tourism development: A viable formula for poverty alleviation in Bergville. *Inkanyiso: Journal of Humanities and Social Sciences, 4*(1), 63–74.

Mulibana, L., & Tshikovhi, N. (2024). Rural entrepreneurship and innovation in BRICS economies: Secondary evidence from rural areas in South Africa. *Sustainability, 16*(6), Article 2408. https://doi.org/10.3390/su16062408

Nassiri-Ansari, T., & Rhule, E. L. M. (2024). Missing in action: A scoping review of gender as the overlooked component in decolonial discourses. *BMJ Global Health, 9*(4), e014235.

Okeke-Uzodike, O. U. E. (2019). Sustainable women's entrepreneurship: A view from two BRICS nations. *Journal of International Women's Studies, 20*(2), 340–358.

Onwuegbuzie, H. N., & Mafimisebi, O. P. (2021). Global relevance of scaling African Indigenous entrepreneurship. *Technological Forecasting and Social Change, 166*, Article 120629. https://doi.org/10.1016/j.techfore.2021.120629

Padilla-Meléndez, A., Plaza-Angulo, J. J., Del-Aguila-Obra, A. R., & Ciruela-Lorenzo, A. M. (2022). Indigenous entrepreneurship: Current issues and future lines. *Entrepreneurship & Regional Development, 34*(1–2), 6–31. https://doi.org/10.1080/08985626.2021.2011962

Parthiban, R., Sun, R., Qureshi, I., & Bandyopadhyay, S. (2024). Empowering rural micro-entrepreneurs through technoficing: A process model for mobilising and developing Indigenous knowledge. *The Journal of Strategic Information Systems, 33*(2), Article 101836. https://doi.org/10.1016/j.jsis.2024.101836

Posel, D. (2025). Who cares? Gender differences in social reproduction and well-being in South Africa. *Gender, Work & Organization*. Advance online publication. https://doi.org/10.1111/gwao.13249

Powoh, T. V. (2016). *Research methods: Quantitative, qualitative, and mixed methods*. Horizons University.

Quagrainie, F. A., & Kabalan, A. (2024). Delivering meaningful poverty alleviation experiences: The role of successful Ghanaian Indigenous women entrepreneurs. In E. L. Eijdenberg, F. Ostertag, L. P. Dana, & H. Onwuegbuzie (Eds.), *Indigenous entrepreneurship in Sub-Saharan Africa: Implications for theory and practice*. Palgrave Macmillan.

Sall, M. A. (2022). Entrepreneurship and poverty alleviation in Africa. *Industrial Policy, 2*(1), 33–37.

Selelo, M. E. (2023). Indigenous entrepreneurship and the Fourth Industrial Revolution in South Africa. *African Journal of Development Studies*, Special Issue *1*, 129–145.

Shash, J. (2021). *Indigenous women and decoloniality: An endless battle?* ResearchGate. https://doi.org/10.13140/RG.2.2.10616.52483

Tenza, T., Mhlongo, L. C., Ncobela, C. N., & Rani, Z. (2024). Contribution of the informal market of village chickens to sustainable livelihoods in KwaZulu-Natal, South Africa. *East African Journal of Biophysical and Computational Sciences, 5*(1), 13–24.

Thorbecke, E., Ouyang, Y., Kedir, A. M., & Odhiambo, S. (2024). The interrelationship amongst growth, inequality and poverty in Africa: Seven studies from five countries. *Journal of African Economies, 33*(Suppl. 1), 1–8. https://doi.org/10.1093/jae/ejae016

Tshamano, N. W., Joshua, M., Terry, M. N., & Lee, K. S. (2023). A new era of entrepreneurship: The transformative potential of African traditional medicine. *Social Sciences, 12*(3), 135–142.

Tshikovhi, N. (2021). A theory of Indigenous entrepreneurship in the modern African state: The case of South Africa and Zimbabwe. *International Journal of African Renaissance Studies: Multi-, Inter- and Transdisciplinarity, 16*(2), 21–60. https://doi.org/10.1080/18186874.2022.2034481

Vorobeva, E. (2022). Intersectionality and minority entrepreneurship: At the crossroad of vulnerability and power. In *Disadvantaged minorities in business* (pp. 225–235). Springer. https://doi.org/10.1007/978-3-030-97079-6_11

Yap, C. S., Keling, W., & Ho, P. L. (2023). Determinants of entrepreneurial performance of rural Indigenous women entrepreneurs in Sarawak, Malaysia. *Gender in Management: An International Journal, 38*(3), 337–356. https://doi.org/10.1108/GM-06-2021-0170

Chapter 7

INDIGENOUS KNOWLEDGE, CLIMATE CHANGE, AND ENTREPRENEURSHIP IN AFRICA

Introduction

Africa faces escalating climate threats as part of the broader global polycrisis. Despite contributing the least to historical greenhouse gas emissions, the continent disproportionately suffers from the environmental and socioeconomic impacts of climate change (Fonjong et al., 2024). However, dominant climate narratives grounded in Western epistemologies often obscure the structural roots of this vulnerability. Terms like "anthropocene" and "anthropogenic climate change" tend to universalize responsibility, concealing the historical role of colonialism, global capitalism, and industrialization, primarily driven by Western Europe and North America, in producing accelerated climate change (Islam et al., 2024). This framing erases accountability and sustains what scholars refer to as climate coloniality, where past and present systems of domination shape both the crisis and proposed solutions (Islam et al., 2024). Consequently, Africa remains acutely vulnerable to the impacts of climate change, a reality shaped by long-standing structural inequalities rooted in colonialism, racism and sexism (Mzimela, 2024). This vulnerability is intensified by the continent's dependence on climate-sensitive economic sectors (Gunaratne et al., 2025).

Across Africa, Indigenous communities are among the most climate-vulnerable groups due to historical and structural injustices (Ebawala Pitiyalage, 2024). Moreover, the interdependence between their cultural identities, natural environment and economic systems exacerbates the devastating impacts of climate change (Dharmasiri et al., 2025; Tirivangasi & Kontinen, 2025). As Karadayı et al. (2025) highlight, communities whose livelihoods depend on agriculture and natural resources in rural settings, many of whom are Indigenous, are among the disproportionately affected. Their vulnerability is not only ecological but deeply political as climate change magnifies existing inequalities stemming from political and economic marginalization, land dispossession, discrimination, and systemic neglect (Dharmasiri et al., 2025). As Islam et al. (2024) note, many Indigenous people are already living in what resembles a dystopian climate future, where their lands, identities, languages, and ways of life continue to be threatened by the legacies of colonialism and extractive development. These lived realities raise questions about the adequacy of mainstream approaches to climate change adaptation.

Climate change adaptation (hereafter, adaptation) is defined by the Intergovernmental Panel on Climate Change (IPCC) as the adjustment of natural and human systems in response to actual or anticipated climate stimuli to moderate harm or capitalize on potential benefits (IPCC, 2021). While widely cited, this definition reflects a Eurocentric, modernist worldview that separates nature from society, framing climate change as an external problem to be managed through technical and bureaucratic interventions (Islam et al., 2024; Sultana, 2022). This framing is therefore problematic as it presents adaptation as a technical and apolitical process, obscuring the colonial and extractive foundations of global climate governance.

Moreover, DeBoom (2022) critiques the way dominant climate narratives, under the guise of urgency, legitimize top-down interventions that often disregard local realities. This form of fraudulent universalism masks the uneven distribution of responsibility and consequence. It enables adaptation strategies that reproduce global hierarchies, create sacrifice zones, and perpetuate extractive logics under the guise of neutrality (DeBoom, 2022). Islam et al. (2024) further caution against the imposition of uniform adaptation models across the Global South. This is because these models erase diverse historical experiences and marginalize Indigenous Knowledge Systems (IKS), framing entire regions as uniformly vulnerable. This flattening of differences opens the door for technocratic solutions by supranational entities that often displace local voices and deepen socio-political marginalization. These approaches prioritize what can be done over what should be done, and by whom, obscuring the central issues of justice, accountability, and agency (DeBoom, 2022).

It is within this context that this chapter affirms the legitimacy of African epistemologies and centers African agency as essential to driving adaptation. We argue that true adaptation must be driven from within, and entrepreneurship, now recognized as a crucial response to climate change, offers a powerful avenue for building resilient economies and livelihoods (Craig et al., 2024; Karadayı et al., 2025). In fact, entrepreneurship is now widely acknowledged as a vital strategy through which Indigenous communities respond to the growing challenges of climate change (Ebawala Pitiyalage, 2024). Dharmasiri et al. (2025) and Tengeh et al. (2022) highlight that within Indigenous communities, entrepreneurship is not a recent or externally imposed concept, but rather a practice deeply ingrained in their cultural values, social systems, and deep relationships with the land. Therefore, Indigenous entrepreneurship is not only a survival mechanism but also a culturally rooted response to climate change (Dharmasiri et al., 2025).

This chapter adopts an Afrocentric and intersectional framework to examine the colonial structures that continue to underpin climate adaptation efforts in Africa. By challenging Eurocentric, deficit-based narratives, this chapter foregrounds the legitimacy and relevance of IKS while illuminating the often-overlooked adaptive capacities embedded within Indigenous entrepreneurial practices. This chapter makes a critical intervention at the nexus of climate adaptation and Indigenous enterprise, an area that has been largely sidelined in scholarship (Dharmasiri et al., 2025). Rather than framing Indigenous communities as passive recipients of adaptation, we position them as knowledge holders and agents of innovation who mobilize IKS and local ingenuity to navigate climate disruptions (Dharmasiri et al., 2025).

Methodologically, the chapter adopts a two-part qualitative approach. It combines a critical review of the literature with empirical data collected in South Africa's uMkhanyakude District Municipality (UKDM). The district is divided into four local municipalities: the Big 5 Hlabisa, Mtubatuba, Jozini, and Umhlabuyalingana (UKDM IDP, 2018/2019). Moreover, UKDM is a predominantly rural region characterized by persistent poverty, climate change-induced water scarcity, and heavy reliance on rain-fed agriculture (Mzimela & Moyo, 2025). These interlinked vulnerabilities make UKDM a critical site for examining how Indigenous entrepreneurs, particularly small-scale farmers, adapt to climate change through enterprise grounded in IKS. Drawing on interviews with Indigenous entrepreneurs, the chapter presents examples of how communities respond to climate change by leveraging IKS. These entrepreneurial ventures challenge the dominance of technocratic, top-down adaptation models and demonstrate the transformative potential of community-led and culturally embedded strategies. In doing so, the study contributes to a growing body of scholarship that calls for decolonizing climate adaptation by centering African epistemologies and restoring Indigenous agency in defining futures. The chapter is structured as follows: it opens by introducing the theoretical lens of Afrocentricity alongside intersectionality to frame the analysis. It then delves into key themes of discussion, such as the shortcomings of mainstream climate adaptation models, the central role of IKS in fostering resilience, and the ways in which entrepreneurship, especially within Indigenous communities in Africa, serves as a culturally grounded response to the climate crisis. It concludes by synthesizing key insights and by arguing for a paradigm shift that places African communities at the center of climate adaptation discourse.

Bridging Afrocentricity and Intersectionality

Afrocentricity, introduced by Molefi Kete Asante, is both a philosophical stance and a methodological framework that places African values, perspectives and interests at the center of analysis (Asante, 2017; Jean-Pierre et al., 2025; Jimoh, 2022; Sakala, 2024). It therefore offers a valuable foundation for reclaiming and rebuilding disrupted African knowledge systems within a decolonial framework (Jimoh, 2022). That is, Afrocentricity plays a crucial role in the decolonization agenda, as epistemological decolonization calls for moving beyond Western-centric thinking and recognizing the value and legitimacy of African ways of knowing (Jimoh, 2022). Afrocentricity emerged as a response to the limitations of Eurocentric frameworks, which have historically misrepresented or ignored the lived realities of African people (Sakala, 2024). It challenges this marginalization by reclaiming African cultural and intellectual agency and positioning Africans as subjects rather than objects in the interpretation of their own realities (Asante, 2000; Sakala, 2024; Sané, 2025). By doing so, Afrocentricity advances a more pluralistic and inclusive epistemic landscape (Jean-Pierre et al., 2025; Sané, 2025; Tirivangasi & Nyahunda, 2025) as it is not an antithesis of Eurocentricity (Jimoh, 2022). According to Jean-Pierre et al. (2025), the theory's core dimensions include recognizing the interconnectedness of all life, nature, and the cosmos; affirming spirituality as central to understanding reality; emphasizing collective identity and shared responsibility; and valuing the affective aspects of lived reality.

Applying an Afrocentric lens is imperative when working within African contexts, as it challenges the imposition of external models and resists the notion that foreign solutions are universally applicable (Ndabaningi et al., 2023). This chapter draws on Afrocentricity to examine the role of IKS, in particular Indigenous entrepreneurship, in climate adaptation. It foregrounds African epistemologies and lived experiences, recognizing Indigenous entrepreneurs as agents of climate resilience. By reclaiming interpretive agency and resisting universalist, technocratic solutions, Afrocentricity creates space for grounded and context-specific responses to climate challenges. It offers an essential corrective to Eurocentric frameworks that continue to dominate scholarly discourse and policy approaches. Ultimately, this approach enables researchers to engage more authentically with African realities and develop solutions that are culturally coherent and community-led (Sané, 2025).

Differently stated, the epistemological contribution of Afrocentricity lies in its affirmation of locally generated knowledge, IKS in this case, which enables the development of adaptation responses that are not only effective but culturally resonant (Sakala, 2024). Rooted in centuries of environmental interaction, Indigenous knowledge in Africa offers historically grounded and socially accepted pathways for navigating climate change (Leal Filho et al., 2025; Mzimela & Moyo, 2024). From this perspective, IKS is not an alternative to Western science but a legitimate and vital source of climate resilience. Indigenous Knowledge offers valuable tools for climate adaptation by enabling communities to apply in-depth understandings of local ecosystems, weather patterns, and environmental changes. This knowledge underpins adaptive practices, such as the selection of climate-resilient crops, livestock breeding strategies, land-use patterns, and disease management techniques tailored to specific ecological contexts (see Leal Filho et al., 2025; Mzimela & Moyo, 2024).

While Afrocentricity provides a valuable foundation for centering African epistemologies and reclaiming agency, it is not without critiques. One limitation lies in its potential to essentialize African identity by assuming a unified or monolithic African experience. Such a perspective can overlook the layered and intersecting identities that shape the lived realities of African people across the continent. To address this, the chapter incorporates Crenshaw's theory of intersectionality, which offers a more nuanced analytical lens to capture the complexities within African communities. Intersectionality highlights the interconnected and mutually reinforcing nature of social identities, including race, class, gender, disability, nationality, and religion, and how these identities influence differential access to power and resources (Yacob-Haliso, 2025). Rather than relying on single-axis explanations like patriarchy, intersectionality emphasizes a "matrix of domination" where structural inequalities are co-constitutive and operate across multiple levels, from the personal to the institutional and global (Yacob-Haliso, 2025). Within the context of climate change, this lens is especially relevant, as Mzimela & Moyo (2025) argue that the interplay of their overlapping identities shapes individuals' adaptive capacities and vulnerabilities. Adaptation strategies that overlook these dynamics risk perpetuating systemic exclusions (Mzimela & Moyo, 2025). For example, the absence of intersectional analysis in climate adaptation policies may unintentionally exacerbate social disparities and fail to meet the diverse needs of vulnerable populations (Ndabaningi et al., 2023).

Therefore, this chapter draws on intersectionality to complement Afrocentricity, ensuring that the diversity within African communities is not only recognized but also analytically foregrounded in discussions of climate adaptation. The point being made is that African identity is not homogenous; the continent encompasses 55 nation-states and a multiplicity of ethnicities, languages, belief systems, traditions, and historical narratives (Yacob-Haliso, 2025). Moreover, Africa is marked by vast diversity in its social structures, economic systems, ecosystems, and livelihoods across different regions (Leal

Filho et al., 2025). While shared ancestry and legacies of anti-colonial struggle foster a sense of collective identity (Jean-Pierre et al., 2025), this solidarity must not obscure internal differences within the continent. Recognizing diversity within African communities is not a denial of unity, but a commitment to more inclusive, context-sensitive scholarship and practice.

By applying this intersectional lens, the chapter examines how social identities, particularly gender and geographical location, influence adaptive entrepreneurial ventures. The point being made is that an Indigenous entrepreneur is never just that; they are always embedded within a broader matrix of identity, which collectively shapes their opportunities, constraints, and adaptive responses. For instance, research shows that men and women often have differing adaptive capacities due to gendered social roles, with women frequently occupying a disadvantaged position in adaptation efforts (Mzimela & Moyo, 2025). However, women, despite bearing the brunt of climate change impacts, have long assumed roles in natural resource stewardship and contribute meaningfully to adaptation strategies (Ndabaningi et al., 2023). Moreover, as highlighted in Chapter Five, Indigenous women are often more likely than men to initiate entrepreneurial ventures (Demong et al., 2023), positioning them as key agents in adaptation. This chapter, therefore, draws on intersectionality to foreground gender and expose the multiple, interwoven systems of advantage and marginalization that shape IE and climate adaptation. By integrating an Afrocentric-intersectional framework, this chapter avoids reducing African identity to a single narrative and instead engages with the diverse, situated experiences of African communities. In doing so, it promotes a more just, responsive, and grounded understanding of climate adaptation in Africa.

Unpacking the Failures of Climate Adaptation

Western frameworks continue to dominate global knowledge production, often dismissing or marginalizing contributions from the Global South (Yacob-Haliso, 2025). This imbalance is rooted in the enduring coloniality of knowledge, which privileges Eurocentric epistemologies while systematically silencing alternative worldviews (Mignolo, 2007; Ndlovu-Gatsheni, 2020, 2018, 2015). Hence, dominant climate adaptation models, constructed mainly by Western actors, are embedded within this epistemic hierarchy. These models tend to prioritize depoliticized, measurable and material outcomes, such as financial losses or technological solutions (Leal Filho et al., 2025; Tirivangasi & Nyahunda, 2025), thereby narrowing the scope of what constitutes legitimate knowledge and action in climate adaptation (Islam et al., 2024). Afrocentricity theory foregrounds the intentional exclusion of

African worldviews and IKSs as part of a broader epistemic apartheid, a global system that ignores and delegitimizes African intellectual traditions and lived experiences. For instance, the IPCC often prioritize technological and market-based adaptation strategies, while overlooking relational, land-based, and community-driven responses rooted in Indigenous ontologies (Dharmasiri et al., 2025). This exclusion is deliberate and reflects a continuation of colonial logics (of extractivism, erasure and dispossession) in contemporary climate governance. As Islam et al. (2024) argue, adaptation strategies risk reproducing the very structures of appropriation, extraction, and ecological violence that characterized imperial expansion, particularly in settler-colonial contexts.

Moreover, climate necropolitics is evident in the way adaptation strategies often extract value and solutions from communities that have contributed the least to climate change, while leaving them exposed to its most severe effects (DeBoom, 2022). These adaptation projects frequently depoliticize vulnerability and frame it as a resource, turning suffering into a site for further exploitation through extractivist mitigation initiatives (DeBoom, 2022). A case in point is the Indigenous Sengwer community in Kenya, whose dispossession under Reduced Emissions from Deforestation and Forest Degradation (REDD+) highlights how adaptation can replicate colonial conservation practices. In this case, the global response to climate change has led to a surge in REDD+ projects aimed at curbing deforestation and promoting carbon sequestration, supported by influential institutions such as the World Bank (Kurang, 2020). However, the implementation of REDD+ has resulted in the forced eviction and dispossession of the Sengwer people, echoing colonial-era conservation practices (Kurang, 2020). Rather than empowering these communities, these projects reflect dominant Western, neoliberal paradigms that exclude local voices and reduce Indigenous peoples to passive observers in their own lands (Kurang, 2020).

Similarly, Lomax et al. (2021) provide a compelling critique of solar mini-grid projects in Kenya's Turkana region, revealing how initiatives framed as climate adaptation can, in practice, reproduce extractive dynamics and exclusion. While these interventions promise renewable energy access for underserved communities, the study shows that high connection fees and market-based tariffs under private models systematically exclude low-income households and those living on the outskirts of villages. This exclusion operates along intersecting lines of gender, with Indigenous women and economically marginalized groups sidelined in land negotiations and community consultations, perpetuating gendered inequalities and reinforcing existing local hierarchies. It is not simply that Indigenous people are excluded, but that particular Indigenous bodies face compounded barriers (Lomax et al.,

2021). Moreover, despite adherence to environmental safeguards on paper, the deployment of hybrid solar-diesel systems has resulted in localized ecological damage, including land degradation and pollution in sensitive areas. Elite capture is also evident, where influential individuals within community energy committees manipulate decision-making processes to prioritize their commercial gain, ultimately undermining equitable access to energy (Lomax et al., 2021).

The exclusion of Indigenous worldviews from mainstream climate adaptation discourse raises serious ethical and epistemological concerns, particularly when Western frameworks are treated as universally applicable. Such approaches often ignore the cultural, spiritual, and socioeconomic realities of Indigenous communities and reinforce portrayals of them as passive victims rather than as holders of IKS with adaptive capacity (Mzimela, 2024). This mischaracterization undermines Indigenous agency, legitimizes state control over ancestral lands, and marginalizes Indigenous governance systems and traditional environmental stewardship practices (Mzimela, 2024). For Indigenous people, adaptation is not a technocratic fix but an extension of centuries of relational living with the land, where nature and human existence are deeply intertwined (Islam et al., 2024). In contrast, settler-colonial narratives tend to frame adaptation as a tool for managing nature in the service of capitalist extraction (Islam et al., 2024).

Afrocentricity presents a critical counterpoint by emphasizing that climate change must be understood within the historical, cultural, and social contexts of African communities (Tirivangasi & Nyahunda, 2025). From this perspective, any meaningful and sustainable adaptation strategy must be embedded in the lived experiences and knowledge systems of those most affected. Addressing climate coloniality demands dismantling the economic, political, and social systems that uphold and reproduce colonial power dynamics (DeBoom, 2022). Afrocentric epistemology rejects the imposition of foreign models and resists the framing of Indigenous knowledge as a mere add-on to Western science. Far from promoting tokenistic inclusion, Afrocentricity challenges the very foundations of global adaptation discourse by asserting the legitimacy and centrality of African thought in shaping climate futures. That is, African systems of knowing should be recognized as central, authoritative, and self-sufficient frameworks for understanding and responding to climate change. A critical interrogation of what constitutes adaptation knowledge, how it is produced, who controls it, and whose knowledge is silenced is essential to confronting climate coloniality (Islam et al., 2024). This includes recognizing and integrating plural forms of knowledge, including Indigenous, non-Western, and marginalized perspectives. Without such epistemological pluralism, adaptation remains a technocratic, limited, and extractive process. In the African context, we cannot accept such limiting and limited approaches, as climate risks and economic vulnerability highlight the urgent need for more inclusive and resilient approaches to advance reparative justice and Indigenous sovereignty (Osabutey & Annoh, 2024).

The Limits of Mainstream Entrepreneurship in Climate Adaptation

Entrepreneurship is increasingly recognized as a vital mechanism for addressing the climate change challenges (Karadayı et al., 2025), particularly in uncertain environments that often foster innovation and new ventures (Ebawala Pitiyalage, 2024). That is, climate disruptions can serve as catalysts for entrepreneurial activities that not only generate profit but also contribute to broader social and environmental well-being (Karadayı et al., 2025). Adaptive entrepreneurs are demonstrating how resource-constrained settings can give rise to innovative, context-specific climate solutions (Ebawala Pitiyalage, 2024). However, as highlighted in chapter five, mainstream entrepreneurship remains entangled in Eurocentric frameworks that obscure its colonial roots, roots shaped historically by slavery and oppression of individuals from the global South. These legacies continue through postcolonial economic policies that reproduce Western-oriented models, often misaligned with the lived realities and knowledge systems of Indigenous communities.

The increasing frequency and severity of climate-related events, such as droughts, floods, and heatwaves, pose serious threats to lives, livelihoods, and economic stability across Africa (Tirivangasi & Nyahunda, 2025). Given that entrepreneurship is a key driver of socioeconomic development at both local and national levels, climate change poses significant structural economic challenges (Beyene et al., 2025). Small and medium-sized enterprises (SMEs), which account for a significant share of employment and Gross Domestic Product (GDP) in many African economies, face heightened exposure to these disruptions. Climate-sensitive sectors like agriculture, fisheries, forestry, and ecotourism, on which many SMEs rely, are particularly vulnerable to erratic rainfall, water shortages, and increasing energy costs, all of which constrain production and weaken supply chains (Osabutey & Annoh, 2024).

While climate change creates opportunities for innovative and adaptive forms of entrepreneurship (Karadayı et al., 2025; Wei et al., 2025), it must be problematized. From an Afrocentric perspective, such narratives can obscure the real vulnerabilities faced by entrepreneurs in African contexts, many of whom operate within informal economies, lack access to capital, and are excluded from formal support systems. These structural barriers intensify their exposure to climate-related disruptions. In this regard, entrepreneurship itself must be problematized: who gets to adapt, with what support, and under which policy frameworks? The growing literature on climate entrepreneurship often prioritizes success stories of innovation while sidelining the everyday realities of adaptation, such as reliance on IKS, communal resource sharing, and subsistence-level adjustments.

Various forms of entrepreneurship have emerged in response to climate-related challenges, including sustainable entrepreneurship, ecopreneurship, environmental entrepreneurship, circular entrepreneurship, social entrepreneurship, and green entrepreneurship (Shahidullah et al., 2023). These models generally promote responsible resource use and are responsive to ecological change (Baba & Fortin-Lefebvre, 2021; Karadayı et al., 2025; Shahidullah et al., 2023). However, IE remains conspicuously absent from these frameworks, even though it holds just as much, if not more, potential for innovation rooted in local realities. This silence is not surprising, however. Many of the other models still draw from Western epistemologies that tend to overlook IKSs developed through generations of living closely with the land and community (Baba & Fortin-Lefebvre, 2021). On the other hand, IE draws from IKS that operate both formally and informally, offering grounded, context-specific strategies for managing climate risks.

Nonetheless, IE should not be treated as peripheral but as central to climate adaptation discourse. Its potential to drive transformative adaptation depends on supportive policy frameworks, institutional recognition, and the inclusion of Indigenous entrepreneurs as both knowledge holders and actors who themselves face the burdens of climate change. Recognizing and investing in IE broadens the conversation beyond Eurocentric frameworks and makes space for plural, locally rooted pathways to resilience. To adequately capture the complexity of entrepreneurship in a changing climate, especially within African contexts, research must shift away from universalist, market-centric frameworks. There is a pressing need to develop context-sensitive, decolonized approaches that recognize the embeddedness of African enterprises within social, ecological, and historical systems. Without this epistemic shift, policies risk reproducing colonial logics, treating African entrepreneurs as recipients of external innovation rather than agents of contextually grounded adaptation.

Climate Change Adaptation Through Indigenous Entrepreneurship

Africans view themselves as active agents rather than passive victims in shaping their responses to climate change. In this context, small-scale farmers in the UKDM are leveraging IKS to adapt to climate change manifestations, in particular drought (Mzimela & Moyo, 2024). Their strategies include harvesting rainwater, digging traditional wells, reusing household water, using buckets for irrigation, and relying on boreholes. To retain soil moisture, many apply manure from livestock, plant trees, leave fields fallow, or use mulch, methods that are especially common among women and households with fewer resources. They also rotate crops, plant hardy or fast-growing varieties, and adjust their planting schedule. These practices are not used in isolation; rather, they work together, demonstrating how farmers blend flexibility and experience to adapt to changing conditions. Long before climate change became a policy focus, these communities had already begun adapting in place. Their deep connection to the land and commitment to sustainability point to the importance of supporting Indigenous approaches as essential tools for adaptation (Mzimela & Moyo, 2024).

In addition to the IKS-based adaptation measures discussed above, farmers shared Indigenous entrepreneurial activities that reflect adaptive ingenuity and cultural continuity in response to drought-related agricultural losses. This move beyond traditional farming into entrepreneurship reflects a broader trend across sub-Saharan Africa (Mzimela & Moyo, 2024). These entrepreneurial ventures reflect both necessity and adaptation, and are influenced by factors such as gender, location, and access to resources. Women in areas like Jozini and Hlabisa discussed selling vegetables and fruits, such as naartjies and mangoes, as well as chickens and handwoven grass mats, an activity observed to be carried out almost exclusively by women, highlighting the gendered nature of adaptation. The weaving and sale of grass mats, in particular, illustrate how cultural heritage and adaptation are intertwined. This activity not only preserves Indigenous craftsmanship but also acts as a low-input, drought-resilient form of income generation. Moreover, this activity embodies a form of "quiet resistance," asserting agency and sustaining livelihoods in ways that remain largely invisible to mainstream adaptation frameworks.

In Mtubatuba, men described cultivating bananas, sugarcane, and cotton, alongside selling livestock, pigs, and fish, as part of their strategies to diversify income in response to increasing climate stress. Among these, Indigenous fishing practices stand out as a unique blend of environmental care and economic initiative. Rooted in long-established ecological knowledge, these fishing methods are guided by customary rules that promote sustainable harvesting and shared responsibility. This approach not only protects natural resources but also enables local fishers to sustain livelihoods through small-scale trade. A similar dynamic is observed among the Luo community in Kenya, as discussed in Chapter Three. Indigenous systems govern fishing on Lake Victoria, dictating when, where, and how to fish, using traditional gear such as traps and spears (Opondo, 2016). These practices help limit overfishing and protect breeding areas, demonstrating how Indigenous Knowledge can serve as a valuable tool for ecological preservation and sustainable livelihoods (Opondo, 2016). Community rules, such as seasonal closures and gear restrictions, act as both conservation strategies and structures for managing local economies. From this foundation, small businesses have emerged in areas like gear-making, fish drying, processing, and local markets. These ventures reflect IE.

IE for climate adaptation is not limited to UKDM. Smallholder farmers in Ethiopia's Tigray region are increasingly adopting Indigenous agroforestry practices to mitigate the growing impacts of climate change, which have disrupted both crop and livestock production (Zeratsiona et al., 2023). What was once a subsistence strategy is now evolving into a more entrepreneurial approach. Trees are generally more resilient than seasonal crops. Hence, farmers have resorted to agroforestry and are now earning an income by selling products such as timber, fruit, fodder, and medicinal plants, making agroforestry a foundation for small-scale business opportunities. This pattern is echoed in other regions. For example, in Niger's Augie district, Indigenous farming communities have responded to extreme weather conditions by diversifying crop production, thereby reinforcing food security through entrepreneurial practices (Dharmasiri et al., 2025). Similarly, small-scale fishers around Lake Malawi have drawn on Indigenous knowledge to expand into agriculture, engage in small businesses, and participate in casual labour within farming and fishing communities, thereby boosting their adaptive capacity in the face of declining fish stocks (Dharmasiri et al., 2025). From an Afrocentric perspective, these entrepreneurial responses cannot be reduced to market logic; they represent relational and place-based responses rooted in Indigenous worldviews that center sustainability, reciprocity, and communal survival. Across these contexts, entrepreneurship functions not only as a coping mechanism but also as a transformative pathway for climate adaptation. That is, in response to growing climate threats, many Indigenous communities draw on IKS to inform entrepreneurial practices that help them manage climate risks and uncertainties, and we refer to these ventures as Indigenous entrepreneurship.

Climate change adaptation is deeply shaped by intersecting inequalities, particularly those related to gender, access to land, and socio-cultural norms. In UKDM, women's engagement in vegetable farming, poultry, and grass mat weaving reflects both adaptive agency and structural constraints, including limited access to land, patriarchal power structures, and marginalization from formal markets. These practices often emerge not from choice alone, but from necessity, with women adapting through low-capital ventures rooted in

Indigenous knowledge and care economies. In contrast, men's participation in sugarcane cultivation, livestock rearing, and fishing reflects broader access to capital, land, and decision-making spaces. Intersectionality helps uncover these complex dynamics, reminding us that adaptation cannot be disentangled from social hierarchies and unequal access to resources. The legacy of colonial and apartheid dispossession in South Africa has compounded these inequities, debunking notions of uniform climate vulnerability (Mzimela & Moyo, 2024).

Literature shows that Indigenous women, often the backbone of rural economies, play crucial roles in climate adaptation through small-scale agriculture, informal trade, and micro-enterprises (Ebawala Pitiyalage, 2024; Gannon et al., 2025). Despite this, their contributions are often overlooked in mainstream entrepreneurship and adaptation discourse. Gender not only shapes adaptation preferences and perceptions but also determines barriers faced within enterprises (Gannon et al., 2025). Although women are disproportionately represented in climate-sensitive sectors, such as agriculture, they remain under-supported in policy and financial ecosystems, despite Africa recording some of the highest rates of female entrepreneurial activity globally (Gannon et al., 2025). This oversight not only undermines effective adaptation but also reinforces gendered exclusion. IE, especially when led by women, fosters inclusive development, enabling marginalized groups to exercise agency and resilience in the face of climate stress (Tirivangasi & Nyahunda, 2025).

Recentering Indigenous Entrepreneurship in Africa's Climate Future

The contemporary global discourse on climate adaptation, while expansive, often perpetuates a Eurocentric bias that marginalizes IKS, particularly in Africa and specifically South Africa's UKDM, where the shadows of colonialism and apartheid linger (Mzimela & Moyo, 2024). However, drawing on IKS for adaptation implies the sensitization and empowerment of indigenous communities (Mzimela & Moyo, 2024). Within climate adaptation, this means recognizing African communities not as passive recipients of global solutions, but as agents of contextually grounded strategies. This is the context within which, in this chapter, we argued that IE is not merely a survival tactic but a potent site of resistance, regeneration, and climate resilience, as Africa stands at the forefront of global climate change impacts, despite having contributed the least to the crisis. Mainstream adaptation narratives continue to overlook the continent's rich Indigenous Knowledge Systems, lived experiences, and adaptive strategies. The dominance of Eurocentric adaptation models, rooted in colonial legacies, technocratic logic, and capitalist rationalities, fails to address the historical and structural injustices that underpin Africa's vulnerability. Against this backdrop, IE, especially when grounded in IKS, emerges as a decolonial strategy that affirms agency and identity and enables context-specific adaptation. Drawing on the case of South Africa's UKDM and other examples across the continent, this chapter demonstrates that IE is deeply rooted in communal values, ecological stewardship, and cultural continuity.

Entrepreneurs in these contexts are not simply market actors; they are custodians of land, culture, and collective survival. Their enterprises reflect intergenerational knowledge, gendered adaptation practices, and spiritual connections to the environment, all of which challenge the homogenizing narratives of vulnerability and technocratic adaptation promoted by global institutions. However, systemic barriers continue to limit the potential of Indigenous entrepreneurs. Women, in particular, bear the disproportionate burden of climate change, while simultaneously driving much of the adaptive innovation in informal and agricultural economies. These realities demand a deliberate rethinking of adaptation policy and research. Therefore, IE must not be treated as a peripheral or alternative approach, but as central to building equitable and locally anchored climate responses.

To do so requires more than recognition; it demands redistribution, repair, and reconfiguration. First, adaptation strategies must be guided by African epistemologies, which resituate African people as agents of their own histories and futures. Second, intersectionality must shape how we understand and respond to climate vulnerability, acknowledging the layered identities that define who adapts, how they adapt, and with what support. Third, Indigenous entrepreneurs must be brought into the center of climate governance, not as beneficiaries, but as co-creators of resilient futures. Finally, this chapter advocates for a shift in how we approach climate adaptation: not as a top-down, technical process imposed from above, but as a relational, place-based practice rooted in history, culture, and justice. IE embodies this shift. It is not only a means of coping with climate shocks, but also a living expression of resistance, knowledge, and hope. By investing in Indigenous entrepreneurial models, revaluing IKS, and centering community leadership, Africa can move toward adaptation strategies that are not only effective but also ethical, equitable, and sustainable. This is not just a call for inclusion but a demand for transformation. This is more so the case because, despite its increasing significance, the intersection of entrepreneurship, Indigeneity, and climate adaptation remains underexplored in academic and policy discourse (Dharmasiri et al., 2025).

Conclusion

Research continues to treat Indigenous communities as recipients of aid rather than as innovators in their own right, often overlooking the ways in which their knowledge systems inform adaptive entrepreneurial responses (Dharmasiri et al., 2025). Even major global reports, such as those produced by the IPCC, largely ignore the potential of entrepreneurship as a transformative force in climate adaptation strategies, particularly in Indigenous and African contexts (Ebawala Pitiyalage, 2024).

However, Africa is one of the most resource-rich continents on Earth, capable of financing and designing its own development trajectories through entrepreneurial solutions (Eshun, 2018). In this context, entrepreneurship should not be seen merely as an economic activity, but as a form of resistance, self-determination, and ecological stewardship. Therefore, now, more than ever, there is a pressing need to “reseed” the continent with Indigenous entrepreneurial approaches that respond to the climate crisis and the enduring legacies of colonial extraction and marginalization. Unlike conventional entrepreneurial models that prioritize profit maximization, IE integrates economic activity with cultural identity, ecological responsibility, and collective well-being (Cheteni & Umejesi, 2024). From this position, Indigenous entrepreneurs are not only economic actors but also stewards of climate resilience, uniquely equipped to identify local challenges and mobilize resources for climate-responsive initiatives (Karadayı et al., 2025).

To fully understand and support IE, there is a pressing need to decolonize climate adaptation discourse and practice. This involves revealing and dismantling colonial ideologies embedded in adaptation policies and practices, which have historically silenced Indigenous agency while promoting technocratic, one-size-fits-all solutions (Sultana, 2022). Moreover, decolonizing climate adaptation requires confronting the legacies and ongoing dynamics of colonialism, capitalism, and development that shape dominant narratives, governance frameworks, and visions of climate-resilient futures (Sultana, 2022). Without addressing these inequities, adaptation efforts risk being dominated by external, often inappropriate, solutions that lack local legitimacy and sustainability (Dharmasiri et al., 2025). Grounding adaptation in African epistemologies and Indigenous agency is therefore not optional; it is central to building climate futures that are just, context-sensitive, and driven from within. In this way, we contribute to a growing body of scholarship that views IE not as an economic outlier, but as a transformative and sustainability-oriented model that redefines adaptation through Afrocentric and decolonial lenses.

References

Asante, M. K. 2017. Afrocentricity and knowledge: Adjusting the narrative lens. *The Thinker,* 74, 24–26. https://hdl.handle.net/10210/466964

Baba, S., & Fortin-Lefebvre, E. 2021. Indigenous entrepreneurship, marginal ontologies, and Sustainable Development Goals. In *Industry, innovation and infrastructure* (pp. 542–553). Springer International Publishing. https://doi.org/10.1007/978-3-319-95873-6_152

Beyene, B., Ngnouwal Eloundou, G., Mokompea Nyamou, J. C., & Gallouj, C. 2025. Energy poverty and entrepreneurship: Evidence from sub-Saharan Africa. *The Bottom Line, 38*(2), 233–252. https://doi.org/10.1108/BL-09-2024-0169

Cheteni, P., & Umejesi, I. 2024. Indigenous knowledge in entrepreneurship and cultural tourism in the rural areas. *Comparative Sociology, 23*(3), 340–359. https://doi.org/10.1163/15691330-bja10110

Craig, C. A., Nochebuena-Evans, L., & Evans, R. 2024. Campgrounds and climate change: An extreme weather event study for nature-based entrepreneurship. *Journal of Business Venturing Insights, 22*, e00477. https://doi.org/10.1016/j.jbvi.2024.e00477

DeBoom, M. J. 2022. Climate coloniality as atmospheric violence: From necropolitics toward planetary mutuality. *Political Geography, 99*, 102786. https://doi.org/10.1016/j.polgeo.2022.102786

Demong, N. A. R., Kassim, E. S., Ibrahim, I., Zamzuri, N. H., Omar, E. N., & Jalil, S. A. 2023. Understanding Indigenous entrepreneurship in Malaysia: A gender perspective. *Journal of Emerging Economies and Islamic Research, 11*(2), 75–93. https://ir.uitm.edu.my/id/eprint/94618

Dharmasiri, I. P., Galappaththi, E. K., Baird, T. D., Bukvic, A., & Rijal, S. 2025. From roots to resilience: Exploring the drivers of Indigenous entrepreneurship for climate adaptation. *Sustainability, 17*(10), 4472. https://doi.org/10.3390/su17104472

Ebawala Pitiyalage, I. P. D. 2024. *Entrepreneurship-based factors to foster climate adaptation among Indigenous communities* (Doctoral dissertation, Virginia Tech).

Eshun, J. P. 2018. From political independence to economic emancipation: Reseeding Africa with entrepreneurship and institutions to accelerate economic development. *Transnational Corporations Review, 10*(3), 244–261. https://doi.org/10.1080/19186444.2017.1287825

Fonjong, L., Matose, F., & Sonnenfeld, D. A. 2024. Climate change in Africa: Impacts, adaptation, and policy responses. *Global Environmental Change, 84*, 102912. https://doi.org/10.1016/j.gloenvcha.2024.102912

Gannon, K. E., Eskander, S. M., Avila-Uribe, A., Castellano, E., Diop, M., & Agol, D. 2025. The role of gender in firm-level climate change adaptation behaviour: Insights from small businesses in Senegal and Kenya. *Climate Risk Management, 48*, 100699. https://doi.org/10.1016/j.crm.2025.100699

Gunaratne, T., Liyanage, S., Punchihewa, C., Badurdeen, S., & Jayathilaka, R. 2025. Unmasking climate vulnerability in Africa: The role of CO_2 and CH_4 emissions on rising temperatures and sea levels. *Humanities and Social Sciences Communications, 12*(1), 1–14. https://doi.org/10.1057/s41599-025-04890-0

IPCC. 2021. *Summary for policymakers.* In V. Masson-Delmotte, P. Zhai, A. Pirani, S. L. Connors, C. Péan, S. Berger, N. Caud, Y. Chen, L. Goldfarb, M. I. Gomis, M. Huang, K. Leitzell, E. Lonnoy, J. B. R. Matthews, T. K. Maycock, T. Waterfield, O. Yelekçi, R. Yu, & B. Zhou (Eds.), *Climate change 2021: The physical science basis. Contribution of*

Working Group I to the Sixth Assessment Report of the Intergovernmental Panel on Climate Change (pp. 3–32). Cambridge University Press.

Islam, F. B., Naylor, L., Bryan, J. E., & Coker, D. J. 2024. Climate coloniality and settler colonialism: Adaptation and Indigenous futurities. *Political Geography, 114*, 103164. https://doi.org/10.1016/j.polgeo.2024.103164

Jean-Pierre, J., Boatswain-Kyte, A., Collins, T., & Ojukwu, E. 2025. Designing Afro-emancipatory qualitative research with and for Black people. *Qualitative Research, 25*(2), 520–542. https://doi.org/10.1177/14687941241264458

Jimoh, A. K. 2022. Decoloniality, epistemological decolonisation and African knowledge practices. *Bodija Journal: A Philosophico-Theological Journal, 12*, 77–96. https://acjol.org/index.php/bodija/article/view/3102/3050

Kurang, A. (2020). *To what extent does REDD+ initiatives reproduce colonial and racist framing of forest conservation in indigenous communities? A case study of the Sengwer people of Kenya* [Working paper]. SOAS University of London. https://www.researchgate.net/publication/358397349

Leal Filho, W., Lütz, J. M., Totin, E., Ayal, D., & Mendy, E. 2025. Obstacles to implementing Indigenous knowledge in climate change adaptation in Africa. *Journal of Environmental Management, 373*, 123678. https://doi.org/10.1016/j.jenvman.2024.123678

Lomax, J., Osborne, M., Aminga, V., Mirumachi, N., & Johnson, O. 2021. Causal pathways in the political economy of climate adaptation: Winners and losers in Turkana, Kenya solar mini-grid projects. *Energy Research & Social Science, 82*, 102296. https://doi.org/10.1016/j.erss.2021.102296.

Maldonado-Torres, N. 2007. On Coloniality of Being: Contributions to the Development of a Concept. *Cultural Studies*, 21 (2–3): 240–270.

Mzimela, J. H. 2025. *An intersectional analysis of the collective efficacy–drought adaptation nexus by small-scale farmers in uMkhanyakude District Municipality, South Africa* (Doctoral dissertation, University of Zululand).

Mzimela, J. H., & Moyo, I. 2024. Indigenous knowledge system and agricultural drought adaptation in the uMkhanyakude District Municipality, KwaZulu-Natal. *Journal of Asian and African Studies*. https://doi.org/10.1177/00219096241275391

Mzimela, J. H., & Moyo, I. 2025. An intersectional analysis of the collective efficacy–drought adaptation nexus among small-scale farmers in uMkhanyakude District Municipality, South Africa. *Environmental Sociology, 1–17*. https://doi.org/10.1080/23251042.2025.2484480

Ndabaningi, A., Tirivangasi, H. M., & Bastiaensen, J. 2023. An Afrocentric intersectional conceptual framework for smallholder microfinance and climate adaptation in Chimanimani, Zimbabwe. In R. R. Shah, B. B. Karki, & L. J. Handayani (Eds.), *The Palgrave handbook of global social problems* (pp. 1–28). Cham: Springer International Publishing.

Ndlovu-Gatsheni, S.J. 2020. *Decolonization, Development and Knowledge in Africa: Turning Over a New Leaf.* London: Routledge.

Ndlovu-Gatsheni, S.J. 2018. *Epistemic Freedom in Africa: Deprovincialization and Decolonization.* London: Routledge.

Ndlovu-Gatsheni, S.J. 2015. Decoloniality in Africa: A Continuing Search for a New World Order. ARAS, 36 (2):22–50.

Opondo, P. 2016. Fisheries as heritage: indigenous methods of fishing and conservation among the Luo fishers of Lake Victoria, Kenya. In A.-M. Deisser and M. Njuguna

(Eds.), Conservation of Natural and Cultural Heritage in Kenya: A Cross-Disciplinary Approach, 200–211. London: UCL Press.

Osabutey, G. O., & Annoh, W. O. 2024. *Climate Change Resilience for Small and Medium-Sized Enterprises (SMEs) in Africa.* Ghana Climate Innovation Centre, Ashesi University. https://ghanacic.ashesi.edu.gh/wp-content/uploads/2024/12/Climate-Change-Resilience-for-Small-and-Medium-sized-Enterprises-SMEs-in-Africa-1.pdf

Sakala, M. 2024. Afrocentric frameworks in recreation and leisure research: A perspective article. *Frontiers in Sports and Active Living, 6*, 1491824. https://doi.org/10.3389/fspor.2024.1491824

Sané, B. M. 2025. Roots of resilience: Unravelling Yacouba Sawadogo's Afrocentric ecological wisdom through qualitative inquiry. *African Studies Review, 68*(1), 31–49. https://www.cambridge.org/core/journals/african-studies-review/article/roots-of-resilience-unraveling-yacouba-sawadogos-afrocentric-ecological-wisdom-through-qualitative-inquiry/A8D92382C655CC0CF32E47E649CDD62B

Shahidullah, A. K. M. 2023. 'Community-based adaptive entrepreneurship' addressing climate and ecosystem changes: Evidence from a riparian area in Bangladesh. *Climate Risk Management, 41*, 100543. https://doi.org/10.1016/j.crm.2023.100543

Sultana, F. 2022. The unbearable heaviness of climate coloniality. *Political Geography, 99*, 102638. https://doi.org/10.1016/j.polgeo.2022.102638

Tengeh, R. K., Ojugbele, H. O., & Ogunlela, O. G. 2022. Towards a theory of indigenous entrepreneurship: A classic? *International Journal of Entrepreneurship and Small Business, 45*(1), 1–15. https://philpapers.org/archive/TENTAT.pdf

Tirivangasi, H. M., & Kontinen, T. 2025. An Afrocentric approach to climate change adaptation: Indigenous seasonal predictors among Ndau people in Chimanimani in Zimbabwe. *Disaster Prevention and Management: An International Journal, 34*(2), 265–278. https://www.emerald.com/insight/content/doi/10.1108/dpm-05-2024-0130/full/html

Tirivangasi, H. M., & Nyahunda, L. (2025). Determinants of transformative climate change adaptation in rural Zimbabwe: An Afrocentric analysis. *Climate Policy, 25*(1), 1–15. https://doi.org/10.1080/14693062.2025.2521486

UKDM IDP. (2018/2019). *Integrated Development Plan (IDP) review 2018/2019.* http://www.ukdm.gov.za/index.php/integrated-development-plan-idp/idp-2018-2019

Wei, S., Boudreaux, C. J., Stenholm, P., Su, Z., & Zhang, Y. (2025). Climate risk and informal entrepreneurship in emerging economies: The moderating effects of entrepreneurs' socio-cognitions. *Journal of Business Research, 199*, 115518. https://doi.org/10.1016/j.jbusres.2025.115518

Yacob-Haliso, O. 2025. African intersectionalities and decolonisation of African women's and gender studies. *History Compass, 23*(3), e70008. https://doi.org/10.1111/hic3.70008

Zeratsion, B. T., Manaye, A., Gufi, Y., Tesfaye, M., Werku, A., & Anjulo, A. 2024. Agroforestry practices for climate change adaptation and livelihood resilience in drylands of Ethiopia. *Forest Science and Technology, 20*(1), 47–57. https://doi.org/10.1080/21580103.2023.2292171

Chapter 8

SUSTAINABLE LIVELIHOODS AND INDIGENOUS ENTREPRENEURSHIP IN AFRICA

Introduction

As discussed in the preceding chapters, Indigenous entrepreneurship (IE) is an expanding field of scholarship (Jurado & Mika, 2023) that refers to entrepreneurial activities and self-employment initiatives grounded in Indigenous knowledge and led by Indigenous communities (Padilla-Meléndez et al., 2022). These ventures generate economic value while preserving cultural identity and reinforcing community-centered ways of life (Cheteni & Umejesi, 2024). Given that IE places Indigenous communities at the center of economic activity, it demonstrates that development can coexist with the protection of cultural values and traditions. Unlike conventional entrepreneurship, which often prioritizes profit maximization, IE is oriented toward broader goals such as self-determination, environmental sustainability, and community empowerment (Cheteni & Umejesi, 2024; Deka & Goswami, 2025; Padilla-Meléndez et al., 2022). It also serves as a means for Indigenous communities to reclaim control over their lands and rebuild socioeconomic systems disrupted by historical marginalization (Peredo et al., 2004).

As such, IE promotes inclusivity and advances sustainable development. A defining feature of IE is its emphasis on collective decision-making rooted in communal values and the responsible use of natural resources. Common activities include farming, fishing, and craft production, carried out in ways that respect ecological limits and cultural heritage (Cheteni & Umejesi, 2024; Padilla-Meléndez et al., 2022). These enterprises are guided by strong environmental stewardship and prioritize long-term sustainability over short-term gains. Furthermore, Indigenous entrepreneurs often cultivate relationships based on mutual respect and social cohesion, extending beyond purely transactional interactions to strengthen community bonds both within and beyond their local contexts (Cheteni & Umejesi, 2024). This orientation reflects a collectivist ethic, where the well-being of the group is prioritized alongside, rather than separate from, individual interests (Ratten & Dana, 2017). In this context, economic activity is not driven solely by profit maximization, as is often assumed in dominant Western models, but is embedded within shared social goals and responsibilities (Ratten & Dana, 2017).

Prominent Indigenous communities include the Māori in New Zealand, the Lakota in the United States, the Maya in Guatemala, and the Aymara in Bolivia (Mohamed Shaffril et al., 2025). In Africa, an estimated 50 million people identify as Indigenous (Leal Filho et al., 2022). In the South African context, this encompasses a wide range of communities, including the Bantu, Griqua, Khoisan, Khoekhoe, Koranna, Nama, Ndebele, San, Sotho, Swazi, Zulu, Venda, Xhosa, Tsonga, and Tswana, among others (Leal Filho et al., 2022). To clarify how these communities are analytically defined, Indigenous peoples are typically characterized by the following six factors: "(1) attachment to ancestral lands and related resources; (2) being descendants from populations inhabiting regions prior to later inhabitants; (3) having distinctive languages and traditions; (4) economic and geographical domination by later inhabitants; (5) maintenance of social-cultural norms; and (6) subsistence economic and institutional arrangements" (Ratten & Dana, 2017, p. 65). For these groups, land and natural resources are not merely economic assets but are deeply embedded in their identities, cultural practices, livelihoods, and overall physical and spiritual well-being (Leal Filho et al., 2022).

In many African contexts, post-colonial development policies continue to reproduce Western-oriented economic models that are often misaligned with the lived realities and knowledge systems of Indigenous communities. As a result, these frameworks have struggled to address persistent challenges; they have had limited success in addressing structural unemployment, uneven development, and ecological vulnerability. This limitation foregrounds the need to engage Indigenous Knowledge Systems (IKS) and IE as alternative, contextually grounded pathways for development. Supporting and expanding Indigenous economies has the potential to accelerate progress toward the Sustainable Development Goals (SDGs) while fostering more inclusive and locally embedded forms of growth (Deka & Goswami, 2025). As Peredo and McLean (2013, p. 593) note, "Indigenous people themselves have turned to entrepreneurial activity as a means of improving their condition." However, realizing this potential depends on creating enabling conditions that allow IE to emerge and thrive within specific socio-cultural and institutional contexts (Deka & Goswami, 2025). Against this backdrop, this study focuses on the Abathwa in Bergville and the Nguni in Ezinqoleni, two Indigenous communities in KwaZulu-Natal, South Africa.

While a growing body of literature highlights the potential of IE to promote inclusive and sustainable development (Cheteni & Umejesi, 2024; Tengeh et al., 2022), there remains a lack of empirical and theoretical work that systematically links IE to broader livelihood frameworks, particularly in the South African context. Existing studies tend to either romanticize Indigenous practices or treat them as isolated economic activities, without sufficiently analyzing how they interact with structural constraints, institutional environments, and livelihood systems. This creates a gap in understanding not only whether IE contributes to sustainable livelihoods, but also how such contributions are produced, negotiated, and limited in practice. This gap not only constrains understanding of Indigenous economic practices but also obscures their potential contribution to sustainable livelihoods. Accordingly, this study examines how IE operates within these communities and the extent to which it supports the development of sustainable livelihoods.

Using empirical evidence from Indigenous entrepreneurs among the Abathwa in Bergville and Nguni communities in Ezinqoleni, the chapter argues that IE contributes to sustainable livelihoods not simply through asset accumulation, but through the relational co-production of livelihood capitals within contexts of structural constraint. In doing so, it advances two key contributions. First, it demonstrates that livelihood assets in Indigenous contexts are not discrete but are mutually constituted through IKS, social relations, and ecological practices. Second, it shows that IE must be understood simultaneously as a site of resilience and as a response to historically produced vulnerability.

The term Abathwa is a Nguni designation for some of the earliest inhabitants of the region, who encountered Bantu-speaking groups during their southward migrations (Wessels, 2012). They are widely regarded as among the first populations to occupy many parts of Africa (Cornhill et al., 2007). Despite this, the Abathwa have been progressively erased from historical narratives, both through forced assimilation into dominant Nguni communities and through the persistent belief that they no longer exist (Francis, 2007, 2009). This erasure has its roots in the late eighteenth and early nineteenth centuries, when colonial regimes dehumanized the Abathwa, portraying them as subhuman and, in some cases, targeting them for extermination (Francis, 2009). Such depictions, at times likening them to vermin, legitimized violence against them, with both colonial settlers and certain Nguni groups implicated in their marginalization (Francis, 2009). In response to sustained persecution, many Abathwa adopted Nguni identities as a strategy for survival (Francis, 2009). Identifying as Zulu, in particular, offered a degree of protection, as the Zulu identity was associated with political strength and collective solidarity (Francis, 2007). This legacy persists, with some individuals continuing to conceal their heritage due to fear of stigma, exclusion, or discrimination (Wessels, 2012).

Today, Abathwa communities are found in areas such as Bergville, in the Okhahlamba Local Municipality, in the foothills of the Drakensberg Mountains. Land in this area is held under a mix of tenure systems, including freehold, the Ingonyama Trust, and private ownership, with approximately 90% of the land currently subject to restitution claims (Okhahlamba Local Municipality, 2022). Socially, Bergville remains largely patriarchal, with entrenched gender roles shaping community life (Mabizela, 2022). A comparable rural context can be observed in Ezinqoleni, situated within the Ray Nkonyeni Local Municipality in the Ugu District of southern KwaZulu-Natal. The name Ezinqoleni reflects the area's historical role as a wagon rest stop during the missionary era. However, it was later altered to Ezingolweni due to pronunciation difficulties, a name that lacks meaning in isiZulu (Yanga et al., 2024). Given its historical significance, the original name is retained in this study. Land use in Ezinqoleni is predominantly characterized by tribal settlements, small-scale holdings, and commercial agriculture (Commonwealth Network, 2024).

In terms of data generation, this was carried out between December 2024 and April 2025. As indicated, the qualitative methodology adopted was grounded in data generation rather than data collection, because the latter implies that Indigenous communities and their knowledges must be mined. Such a research practice must neither be accepted nor tolerated, as it sustains colonial tendencies and practices that should have no place in society (Alcoff, 1991; Tuhiwai-Smith, 2012). For these reasons, semi-structured interviews with Indigenous entrepreneurs were conducted to generate data around the concept and practice of IE and the meaning and experiences attached to their social existence (Creswell & Creswell, 2005, 2018). Indigenous entrepreneurs were purposively selected from the study communities based on their active engagement in IE. This sampling strategy ensured the inclusion of participants whose enterprises are embedded in IKS and reflect the socio-cultural and ecological realities of their communities. The selected participants were involved in a diverse range of Indigenous enterprises, including traditional agriculture, small-scale livestock farming such as sheep, pigs, and cattle, poultry production, traditional healing practices, beadwork, arts and crafts, grass mat weaving and trading, home-based healthcare services, fishing, as well as the production and performance of traditional music and dance.

This diversity of livelihood activities was critical in capturing the breadth of IE and its role in sustaining livelihoods across different sectors. It also allowed for a nuanced understanding of how these enterprises draw on local resources, knowledge systems, and social networks to generate economic and non-economic value. In total, 19 Indigenous entrepreneurs participated in the data-generation process. This sample size is appropriate for qualitative data generation, in which the emphasis is on depth, richness, and contextual understanding rather than statistical representativeness (Creswell & Creswell, 2005, 2018). Smaller, purposively selected samples are particularly effective at generating detailed, context-specific insights and uncovering participants' meanings and experiences within their lived environments (Creswell & Creswell, 2005, 2018). In this study, the sample enabled an in-depth exploration of how IE operates within specific community contexts and contributes to sustainable livelihoods. The data were analyzed using thematic analysis, which involved familiarization with the data, coding, theme development, and iterative refinement of themes to capture patterns in livelihood assets, strategies, and outcomes.

Thus, drawing on empirical research on selected Indigenous entrepreneurs among the Abathwa and Nguni in South Africa, this chapter demonstrates how IE activities are embedded in community-oriented values that prioritize care, reciprocity, and collective well-being over narrow profit maximization. It argues that these alternative entrepreneurial models challenge dominant market-driven paradigms by fostering inclusive development, strengthening local economies, and reinforcing social cohesion. The chapter highlights the potential of community-centered entrepreneurship to generate sustainable and locally grounded development outcomes. This raises the following questions: How and why does IE lead to sustainable livelihoods? To take this discussion forward, the next section engages more directly with the concept of sustainable livelihoods, positioning it as the paper's central theoretical axis. The Sustainable Livelihood Framework (SLF) is particularly useful in this regard, as it provides the lens for understanding how different forms of capital, which include human, social, natural, financial, and physical, interact within specific vulnerability contexts and institutional environments to shape livelihood strategies and outcomes (Chambers & Conway, 1991; Scoones, 1998). The following section discusses how IE is embedded in social relations, ecological care, and IKS. The intention is to demonstrate that IE does not operate in isolation but is deeply relational, drawing on networks of reciprocity, collective decision-making, and shared cultural values. The final section of the chapter emphasizes that IE offers a viable pathway for locally grounded and sustainable development, particularly in contexts where conventional, externally driven models have proven inadequate or exclusionary.

Theoretical Framing: Rethinking Livelihoods Through Indigenous Entrepreneurship

To theoretically frame how IE leads to sustainable livelihoods in Africa, a useful starting point is to engage with the notion of sustainable livelihoods itself. First, it is necessary to define what a livelihood is. In simple terms, a livelihood is a means of making a living (Amghani et al., 2025; Natarajan et al., 2022). It is constituted of capabilities, assets such as material and social resources, and activities required for making a living (Amghani et al., 2025; Chambers & Conway, 1991). A livelihood is deemed sustainable when it can withstand and recover from stresses and shocks, while maintaining or improving its capabilities and assets over time, without degrading the natural resource base on which it depends (Chambers & Conway, 1991). In this regard, sustainable livelihoods refer to the resources, assets, activities, and strategies that individuals and communities rely on to sustain their living. These livelihoods are shaped by different forms of capital, including human, natural, financial, physical, and social capital (Chambers & Conway, 1991; Scoones, 1998). To better understand how these elements interact, SLF was developed as an analytical tool for examining how people secure and sustain their livelihoods (Scoones, 1998). SLF comprises five interrelated components: livelihood assets, the vulnerability context, transforming structures and processes, livelihood strategies, and livelihood outcomes (Scoones, 1998). These elements provide a holistic lens for analyzing how individuals and communities construct, sustain, and adapt their livelihoods within changing socioeconomic and environmental conditions (Chambers & Conway, 1991; Scoones, 1998).

Livelihood assets are the range of resources and capabilities that people depend on to make a living. These assets are typically categorized into different forms of capital, such as human, natural, financial, physical, and social, and can be both tangible and intangible (Chambers & Conway, 1991). Tangible assets include material resources such as food stocks, savings, land, livestock, trees, tools, and infrastructure. Intangible assets, by contrast, encompass entitlements and opportunities, including access to employment, education, healthcare, information, and social support networks. Crucially, access to these assets is often uneven and shaped by broader structural inequalities, which in turn influence the range of livelihood options available to different groups (Chambers & Conway, 1991; Scoones, 1998). The vulnerability context captures the external environment within which livelihoods are pursued (Afrin & Islam, 2021). This includes shocks such as droughts, floods, and economic crises; long-term trends such as climate change, population growth, and market transformations; and seasonal variations that affect resource availability and income flows. These factors can either constrain or reshape livelihood opportunities and often disproportionately affect marginalized communities with limited adaptive capacity. Understanding the vulnerability context is therefore essential for assessing the resilience and sustainability of livelihoods (Chambers & Conway, 1991; Scoones, 1998).

Transforming structures and processes refers to the institutional, policy, legislation, and socio-cultural frameworks that mediate access to assets and shape livelihood strategies (Scoones, 1998). Simply put, livelihoods are accessed through institutional and social frameworks (Amghani et al., 2025). Policies, institutions, and processes can grant or block access to assets (Trang & Loc, 2021). These include formal institutions such as governments, legal systems, and markets, as well as informal norms, customs, and power relations within communities. For example, land tenure systems, gender norms, and regulatory policies can either enable or restrict access to resources and opportunities. These structures and processes are not neutral; they often reflect existing power dynamics and can reinforce or challenge inequalities in access to livelihood assets (Chambers & Conway, 1991; Scoones, 1998). Livelihood strategies are the diverse set of activities and choices that individuals and households undertake to achieve their livelihood goals (Fischer et al., 2021). These strategies are typically adaptive and may involve combining multiple activities such as farming, trading, fishing, wage labour, or small-scale enterprises to spread risk and enhance resilience. The selection of strategies is influenced by the assets available, the vulnerability context, and the prevailing institutional environment (Chambers & Conway, 1991).

Livelihood outcomes are self-defined and represent the results of these strategies. These may include improved income, food security, reduced vulnerability, enhanced well-being, and the sustainable use of natural resources. Importantly, outcomes are not solely economic; they also encompass social and environmental dimensions of sustainability. The SLF emphasizes that positive livelihood outcomes depend on the effective interaction of all components, highlighting the dynamic and interconnected nature of livelihood systems (Chambers & Conway, 1991; Scoones, 1998). The value of the SLF lies in its ability to provide a holistic understanding of the complex and interconnected factors that influence livelihoods. It enables analysis of how access to resources affects livelihood options and outcomes, as well as how institutional and policy environments shape these processes (Chambers & Conway, 1991; Scoones, 1998).

The SLF embodies an epistemological viewpoint that values indigenous knowledge, elevates local voices, and 'puts the last first' (Natarajan et al., 2022). For the present purposes, the SLF is utilized to analyze how IE contributes to sustainable livelihoods in selected South African Indigenous communities by demonstrating that IE operates across all key dimensions of the SLF in ways that strengthen livelihood resilience and sustainability. As such, the question of how IE enhances livelihood assets by mobilizing and reinforcing multiple forms of capital is engaged. Linked to this is how IE builds human capital by transmitting IKS and skills, strengthening social capital through community networks and collective decision-making, and supporting natural capital through practices grounded in environmental stewardship. Using the SLF, we answer the question of how IE generates financial and physical capital through locally rooted economic opportunities and improved access to resources and infrastructure. Also engaged is the issue of how IE reduces vulnerability by diversifying income sources and embedding economic activities within local ecological and cultural systems to withstand shocks such as climate variability, market fluctuations, and socioeconomic instability. Linked to this is the question of how IE enables the development of diversified, context-specific livelihood strategies that are adaptive, sustainable, and aligned with local ecological and socio-cultural realities, while supporting long-term, self-determined development.

Indigenous Entrepreneurship and the Generation of Financial Capital for Sustainable Livelihoods

From natural capital to financial capital: iks and value creation

This section examines how IE contributes to sustainable livelihoods by generating financial capital through the deployment of SLF. We demonstrate how Indigenous entrepreneurs mobilize locally available resources, Indigenous knowledge, and social capital to create income-generating activities that are both economically viable and ecologically sustainable. That is, IE facilitates the conversion of resources into financial capital through context-specific livelihood strategies. As Soma et al. (2022) note, access to and control over assets determine livelihoods. In both Bergville and Ezinqoleni, natural resources such as *imizi* (grass), livestock, plants, and locally available materials are transformed into economic value through IKS. For example, *imizi* grass harvested along riverbanks is used to produce grass mats, with Indigenous ecological knowledge guiding sustainable harvesting practices and weaving techniques. This ensures that the resource base is not depleted but maintained, while simultaneously generating income. Similarly, fishing along the uThukela River, including the harvesting of species such as yellowfish, relies on detailed knowledge of seasonal cycles and ecological conditions. These low-cost, knowledge-based practices enable the sale or exchange of products in local markets while maintaining the resource base. Small-scale livestock farming, including cattle and sheep, provides another example of how IE facilitates the conversion of natural capital into financial capital within SLF. In this context, natural capital takes the form of land, grazing resources, and locally adapted livestock breeds (such as Blackhead sheep, among Abathwa communities). Through the application of IKS, particularly in breeding, feeding, and animal care practices, these resources are managed in ways that are both cost-effective and environmentally attuned. This enables the production of multiple outputs, for example, livestock can provide meat, milk, and hides, which can be sold to generate income. Importantly, this activity reflects not only the direct transformation of natural resources into financial capital but also a process of value multiplication, where a single asset base supports diverse income streams. In doing so, small-scale livestock farming strengthens financial capital while enhancing livelihood resilience through diversification. In SLF terms, these examples reflect the effective mobilization of existing livelihood assets within specific ecological and socio-cultural contexts (Chambers & Conway, 1991; Scoones, 1998).

Low-input Systems and the Politics of Cost Reduction

Building on this, IKS plays a critical role in reducing production costs, thereby strengthening income generation. Income generated from the sale of these products enables households to meet basic needs, reinvest in their activities, and improve their overall livelihood security. Indigenous entrepreneurs rely on locally embedded practices, such as the use of Indigenous livestock breeds and traditional feeding systems, which minimize dependence on costly external inputs. In livestock and poultry production, this reduces reliance on commercial feed and veterinary services, while in crafts and fishing, locally available materials and techniques lower operational costs.[1] As a result, entrepreneurs retain a greater share of their earnings, enhancing financial capital. In SLF terms, this reflects an effective conversion of natural and human capital into financial capital through context-specific livelihood strategies. Unlike mainstream economic models that often depend on external inputs and extractive practices, IE relies on accessible, locally available resources, reducing costs and increasing economic autonomy (Dana, 2015). However, it is also important to note that limited dependence on external markets also points to restricted access to formal inputs, credit, and institutional support.

Moreover, IE generates financial capital through forms of market engagement that are closely aligned with local demand and cultural contexts. In both study areas, economic activities respond directly to community needs. Traditional healing practices, for instance, address health needs that are not always fully met by formal biomedical systems, thereby creating a stable and culturally embedded market. Similarly, products such as beadwork, grass mats, and other crafts are produced for both local consumption and niche markets that value cultural authenticity. In this way, IE leverages cultural knowledge to create economically viable products that are relevant and competitive within specific social contexts (Cheteni & Umejesi, 2024; Tengeh et al., 2022).

Livelihood Diversification and Resilience in Practice

Furthermore, IE supports livelihood diversification, a central strategy within the SLF. Households in both study sites often engage in multiple economic activities, such as crop farming, livestock farming, and poultry farming. This shows that Indigenous entrepreneurs do not depend on a single income source, but construct livelihood portfolios from available natural, human, social, and cultural assets. This diversification spreads risk and stabilizes income flows, reducing vulnerability to environmental shocks, market fluctuations, and seasonal variability (Scoones, 1998; Chambers & Conway, 1991). Livelihood diversification is a norm (Mahama & Maharjan, 2019). As such, diversified livelihood portfolios represent a key mechanism through which IE enhances livelihood security. Closely linked to this is the role of informal and localized markets in enabling participation. Many Indigenous entrepreneurs operate within informal and community-based markets, where goods and services are exchanged through local networks, social relationships, and informal trading spaces. This reduces barriers to entry, such as high transport costs or strict regulatory requirements, making it easier for Indigenous entrepreneurs to participate in economic activities and generate income (Dana, 2015). At the same time, reliance on informal markets reflects exclusion from formal economic systems. In both study sites, such localized market systems enable broader participation in economic activities, particularly for individuals who may be excluded from formal markets. Importantly, income generated through these activities tends to circulate within the community, supporting other local enterprises and reinforcing local economic systems (Tengeh et al., 2022).

Importantly, these forms of market engagement are not purely profit-driven but are embedded in social and cultural systems that prioritize sustainability and collective well-being. This ensures that financial capital is generated in ways that do not undermine social cohesion or environmental integrity (Mazzocchi, 2020). In SLF terms, therefore, IE demonstrates how financial capital is built through the effective mobilization of human, natural, and social (networks and trust) capital. As Carney (1999) argues, no single category of assets is sufficient to secure positive livelihood outcomes, and the range of livelihood options expands with access to multiple forms of capital (Haidar, 2009). At the same time, livelihood construction involves trade-offs, as assets are combined and substituted in context-specific ways (Simiyu, 2012). Based on aligning production with local needs, reducing dependence on external inputs, and fostering diversified and accessible markets, IKS-based enterprises create sustainable and resilient pathways for income generation within Indigenous communities.

Indigenous Entrepreneurship and Strengthening of Human, Social, and Natural Capital

In this section, we demonstrate how IE, through the transmission and application of IKS, strengthens human, social, and natural capital as seen through SLF, and how the interaction of these forms of capital supports sustainable livelihood outcomes. We further argue that, in Indigenous contexts, these forms of capital are not discrete but relationally co-produced through everyday economic practices.

Intergenerational Knowledge and the Strengthening of Human Capital

First, IE strengthens human capital through the intergenerational transmission of knowledge, skills, and competencies. Within the SLF, human capital refers to the skills, knowledge, education, health, and labor that enable individuals to pursue livelihood strategies and achieve positive outcomes (Goh, 2012). It is also fundamental to the activation and utilization of other forms of capital; without it, the effective use of these assets is limited (Carney, 1999). Activities such as mat-making, farming, and resource harvesting function not only as income-generating practices but also as pedagogical spaces. Learning occurs through participation, for example, working alongside adults during grass mat weaving, where skills are demonstrated, observed, and gradually mastered over time. In this sense, the act of making becomes the curriculum, with knowledge embedded in practice rather than abstract instruction. Through these embodied and relational processes, Indigenous knowledge, technical skills, and cultural practices are reproduced across generations. This form of pedagogy is land-based, experiential, and situated, unfolding within everyday life. As a result, knowledge remains context-specific, responsive, and adaptable to changing conditions (Padilla-Meléndez et al., 2022; Mazzocchi, 2020). This point is amplified by Senanayake (2006, p. 87), who correctly observed that IKS is "rooted to a particular place and set of experiences and generated by people living in those places," meaning that attempts to abstract or transfer it across contexts risk dislocating its meaning and function.

Relational Networks and the Co-production of Social Capital

Second, this process of knowledge transmission simultaneously reinforces social capital, understood as the networks, norms, and relationships that enable collective action and coordination toward shared livelihood goals (Mersha & Van Laerhoven, 2016). As skills and practices are shared within families and across community networks, relationships of trust, cooperation, and reciprocity are strengthened. In this context, IE is less individualized and more relational, often organized around shared decision-making and communal participation rather than competitive accumulation. For example, women engaged in poultry production in Ezinqoleni frequently operate as cooperative groups, while livestock farmers support one another through practices such as *ukusisa,* lending animals as seed capital to help households establish or rebuild herds. This fosters strong social bonds and mutual accountability, enabling communities to coordinate activities, share resources, and support one another in times of need (Tengeh et al., 2022; Cheteni & Umejesi, 2024). In SLF terms, these networks enhance resilience by providing social safety nets and facilitating collective responses to challenges (Scoones, 1998). However, it is also important to recognize that while these strong local networks enable inclusion at the community level, they may not always translate into access to wider markets or formal institutional support, highlighting the uneven scaling of social capital across different economic contexts.

Environmental Stewardship and the Regeneration of Natural Capital

Third, IE contributes to the maintenance and regeneration of natural capital through practices grounded in environmental stewardship. The knowledge transmitted within these systems encompasses detailed understandings of ecosystems, seasonal cycles, and sustainable harvesting methods, which directly shape how resources are accessed and used. In Bergville and Ezinqoleni, this is evident in activities such as the controlled harvesting of *imizi*, where Indigenous knowledge of regeneration rates and seasonal timing ensures that extraction remains within ecological limits. Similarly, traditional healing practices depend on locally sourced medicinal plants, with knowledge systems guiding what species to harvest, when, and in what quantities to prevent depletion. Crucially, these practices are informed by normative obligations to maintain ecological balance and safeguard resources for future generations. This alignment between economic activity and ecological limits is further reflected in Indigenous farming systems, where practices such as mixed cropping and seasonal timing contribute to soil conservation, biodiversity maintenance, and climate resilience.

Such practices reflect a relational approach to the environment, where economic activities are aligned with ecological limits and long-term sustainability (Akbar et al., 2020; Dana, 2015). Indigenous communities have long maintained a symbiotic relationship with the natural environment (Deka & Goswami, 2025). In this sense, IE does not simply draw on natural capital but actively reproduces it through context-specific management strategies. However, this relationship should not be romanticized. While these practices support sustainability, they are also shaped by structural constraints, including limited access to land, environmental degradation, and pressures from external market systems, all of which may undermine long-standing stewardship practices. Importantly, these three forms of capital, human, social, and natural, are deeply interconnected. Strengthening human capital through IKS enables the development of social capital grounded in shared practices and collective governance. In turn, strong social institutions support the sustainable management of natural resources. This integrated system reduces vulnerability and enhances the sustainability of livelihoods by ensuring that economic activities do not undermine the environmental and social foundations on which they depend (Chambers & Conway, 1991; Scoones, 1998). These findings suggest that, in Indigenous contexts, capital cannot be understood as discrete categories but must be seen as relational and mutually reinforcing.

At the same time, IE also contributes to the development of physical capital, though in ways that are modest and closely tied to local realities. In Bergville and Ezinqoleni, this can be seen in the use of simple, locally made tools, such as hand-held implements for harvesting *imizi,* simple frames to assist hand weaving of grass mats, basic fishing equipment, and small livestock enclosures built from available materials. Many activities, including beadwork and traditional healing, are also carried out in home-based or shared community spaces. This keeps costs low and allows people to stay close to both their resources and local markets. Together, these forms of physical capital support production without placing heavy financial or environmental strain on households. Over time, the income from these activities begins to make a difference. Money earned from selling mats, livestock, or fish is often reinvested, whether to buy additional tools, expand herds, improve storage, or upgrade parts of the home. In some cases, people work together, as seen in group-based poultry projects, where resources are pooled to build shared shelters or improve feeding systems. In this way, physical capital develops gradually, shaped by what people are able to earn and reinvest. Through the prism of SLF, IE generates financial and physical capital not through large-scale industrial inputs, but through the strategic use of local resources, knowledge, and social organization. This approach fosters sustainable economic development by enhancing livelihood assets, reducing vulnerability, and ensuring that resource use remains in harmony with environmental systems (Dana, 2015).

Navigating Vulnerability: Assets, Structures, and the Limits of Indigenous Entrepreneurship

It is evident that IE contributes to the strengthening of capitals; however, these processes occur within a context of constrained vulnerability that both enables and limits livelihood strategies. Within the SLF, this context is constituted by the interaction of environmental shocks, long-term structural inequalities, and institutional conditions that influence access to assets and the viability of livelihood strategies (Afrin & Islam, 2021; Fischer et al., 2021). In both study sites, structural inequality remains a defining feature of this context. Insecure land tenure, rooted in historical dispossession and ongoing restitution processes, continues to restrict access to natural capital and discourages long-term investment in agriculture. These constraints are further compounded by gendered inequalities, where women entrepreneurs face limited access to finance, networks, and recognition, while also carrying disproportionate unpaid care responsibilities. Similarly, the marginalization of youth reflects restricted access to skills, resources, and institutional support necessary for enterprise development. Taken together, these dynamics demonstrate that access to livelihood assets is not neutral but deeply shaped by entrenched power relations.

These structural conditions intersect with institutional and market constraints that directly affect the accumulation of financial and physical capital. Restricted access to broader networks, for example, limits Indigenous entrepreneurs' ability to engage with formal financial institutions (Ratten & Dana, 2017). Consequently, the absence of formal financing options compels many to rely on informal funding mechanisms or to diversify their livelihood activities in order to remain viable. At the same time, inadequate infrastructure, including unreliable water supply and the geographic isolation of rural communities, restricts access to markets, services, and support systems (Selelo, 2023). The closure of grant collection centers has further intensified this isolation by removing key spaces for social interaction and informal trade. As a result, many entrepreneurs operate within informal markets characterized by unstable demand, non-payment, and limited bargaining power, conditions that constrain income generation and weaken the social capital that underpins cooperative enterprise. Environmental pressures add a further dimension to this vulnerability context. Prolonged drought has reduced the availability of key natural resources, including reeds used for mat weaving and grazing for livestock, directly undermining natural capital and the livelihood strategies that depend on it. Importantly, such environmental shocks do not operate in isolation but amplify existing socioeconomic vulnerabilities, reinforcing the interconnected nature of the SLF components.

However, these conditions do not simply determine outcomes; they are actively mediated through what the SLF terms transforming structures and processes. In these communities, access to land, tools, and production spaces is not primarily governed by formal institutions but is organized through customary norms, social networks, and collective arrangements. These informal systems regulate who can access resources, under what conditions, and for what purposes, thereby shaping how livelihoods are organized in practice. In this sense, transforming structures do not merely reflect vulnerability but actively shape how it is navigated. For example, while insecure land tenure limits formal ownership, customary systems of access often enable continued use of land for farming and grazing. Similarly, although formal financial institutions remain largely inaccessible, savings groups, social grants, and reciprocal exchanges provide alternative mechanisms for mobilizing capital and sustaining enterprise activity. These arrangements allow Indigenous entrepreneurs to operate within constrained environments by drawing on locally embedded systems of support.

At the same time, these structures are not inherently enabling. While they facilitate participation and inclusion at the local level, they may also reproduce existing power relations and limit access to broader economic opportunities. Reliance on informal markets, for instance, lowers barriers to entry but also exposes entrepreneurs to instability, exploitation, and limited growth potential. In this way, transforming structures simultaneously mitigate and reproduce vulnerability, highlighting their dual and often contradictory role within the SLF. These dynamics help explain the form that IE takes in these contexts. Practices such as livelihood diversification, reliance on low-cost production methods, and the use of locally available resources reflect not only Indigenous knowledge and adaptive capacity but also responses to structural exclusion. The reliance on informal exchange systems and community-based networks, for example, reflects both the strength of local social capital and the absence of accessible formal institutions.

Taken together, these findings point to a central tension within IE. On the one hand, Indigenous entrepreneurs demonstrate considerable adaptability by mobilizing available assets to sustain livelihoods under challenging conditions. On the other hand, this adaptability is exercised within a broader system that continues to marginalize Indigenous economic practices. IE should therefore be understood not only as a site of resilience but also as a response to historically and structurally produced vulnerability. Viewed through the SLF, this interplay between vulnerability and transforming structures highlights that livelihood outcomes are shaped not only by the availability of assets but by the broader political and economic context that governs access to them. In Bergville and Ezinqoleni, sustainable livelihoods emerge not from the absence of constraint, but from the capacity of Indigenous entrepreneurs to navigate, negotiate, and rework these constraints through context-specific strategies.

Importantly, these processes are embedded within transforming structures and processes that are often informal and community -driven. Access to tools, land, and workspaces is usually negotiated through social networks, customary practices, and collective arrangements rather than formal institutions. This makes it easier for people to participate, but it also points to the limits they face in accessing formal infrastructure or support. Moreover, Indigenous systems of resource governance regulate the harvesting of materials, ensuring regeneration and long-term sustainability. This not only safeguards natural capital but also stabilizes the economic activities that depend on it. In doing so, IE reduces vulnerability by minimizing exposure to external shocks such as fluctuating input costs or environmental degradation. Furthermore, the alignment of livelihood strategies with local ecological conditions ensures that economic activities remain sustainable over time. The use of IKS enables communities to balance production with conservation, demonstrating that income generation need not come at the expense of environmental integrity. As Akbar et al. (2020) note, such locally grounded solutions are context-specific and responsive to both socioeconomic and ecological challenges.

What also needs amplification is that, from an SLF perspective, the evidence from the two Indigenous communities in KwaZulu-Natal shows that IE is oriented less toward individual profit maximization and more toward collective well-being and livelihood sustainability. While income generation remains important, it is embedded within a broader objective of community development, a pattern consistent with findings in other contexts (Colbourne et al., 2023; Colbourne, 2017). This reorientation shifts the emphasis from individual accumulation of financial capital to the strengthening of shared livelihood assets, particularly social and human capital. In these communities, the benefits of entrepreneurial activity are distributed across the collective, ensuring that gains are not concentrated in the hands of a few individuals. As a result, the success of Indigenous entrepreneurs is inseparable from the well-being of the community as a whole. This reflects a livelihood system in which financial capital is generated and circulated in ways that reinforce social cohesion, mutual support, and collective resilience. Within the SLF, this illustrates how livelihood outcomes extend beyond income to include enhanced social relations, reduced vulnerability, and improved community well-being.

Additionally, the foundations of IE in these contexts are deeply rooted in ancestral knowledge, spiritual relationships with the environment, and systems of reciprocity. These elements shape transforming structures and processes by embedding economic activities within culturally governed norms of sharing, cooperation, and environmental stewardship. Entrepreneurships are therefore not isolated economic units but are relationally embedded within networks of care, labor exchange, and ecological responsibility (Mazzocchi, 2020). This collective orientation also contributes to sustainability. Based on prioritizing community welfare and ecological balance, IE aligns livelihood strategies with long-term resilience rather than short-term gains. In SLF terms, this enhances the durability of livelihood outcomes by ensuring that economic activities do not undermine the natural resource base or social fabric upon which they depend. Thus, in these communities, IE demonstrates that sustainable livelihoods are best understood not simply in terms of financial returns but in terms of the capacity to sustain and strengthen the collective systems that support life and well-being.

Conclusion

In this conclusion, it is necessary to refer to the overarching question we set out to answer: whether and how IE leads to sustainable livelihoods that can provide a foundation for inclusive socioeconomic development in Africa. The analysis demonstrates that, when viewed through the lens of the SLF, IE does indeed provide a viable and contextually grounded pathway toward such outcomes. The findings show that IE operates through locally embedded practices that mobilize existing livelihood assets, particularly natural, human, and social capital, within specific ecological and socio-cultural contexts (Cheteni & Umejesi, 2024). Through IKS, these assets are transformed into financial and physical capital in ways that are both sustainable and adaptive (Dana, 2015). For instance, the use of locally available resources, combined with culturally informed production and governance systems, reduces dependence on external inputs while enhancing economic autonomy and resilience. In SLF terms, this reflects an effective and sustainable conversion of livelihood assets into improved livelihood outcomes (Chambers & Conway, 1991).

Importantly, these processes do not occur in isolation. IE is deeply relational, rooted in systems of reciprocity, collective decision-making, and ecological stewardship (Leal Filho et al., 2022). As a result, it strengthens social cohesion, supports knowledge transmission, and ensures the sustainable management of natural resources. These interconnected dynamics contribute to diversified and resilient livelihood strategies, which in turn lead to outcomes such as improved income, reduced vulnerability, and enhanced community well-being. Crucially, these outcomes extend beyond individual gain to encompass broader, collective benefits, thereby aligning with the principles of inclusive socioeconomic development (Dana, 2015). The chapter underscores that IE offers more than an alternative economic model because it represents a holistic development paradigm that integrates economic activity with social and environmental sustainability (Mazzocchi, 2020). This is particularly significant in African contexts, where conventional, externally driven development models have often failed to address local realities or have exacerbated inequalities. At the same time, the study highlights that these livelihood processes unfold within conditions of structural constraint. Insecure land tenure, gendered inequalities, limited access to finance and markets, and environmental pressures such as drought continue to shape and limit the possibilities of Indigenous enterprise. IE therefore emerges not only as a site of resilience, but also as a response to historically produced vulnerability. Recognizing this tension is critical, as it shifts the interpretation of IE from a purely successful alternative model to one that reflects both adaptive capacity and structural exclusion.

Although the empirical insights are drawn from selected Indigenous communities in South Africa, their implications resonate more broadly across the African continent. Many communities share similar socioeconomic conditions, cultural frameworks, and ecological dependencies, suggesting that the principles underpinning IE have wider applicability. As such, IE holds considerable potential for informing development strategies that are locally grounded, culturally relevant, and sustainable (Dana, 2015). For this reason, it can be stated that IE, when understood through the SLF, does lead to sustainable livelihoods and has the capacity to contribute meaningfully to inclusive socioeconomic development in South Africa and beyond. All this highlights the need to recognize, support, and integrate Indigenous entrepreneurial practices into broader development policy frameworks to foster resilient and equitable futures. Against this backdrop, the chapter's practical implications become clear. If IE is to realize its potential as a pathway for inclusive and sustainable development, greater attention must be given to the structural barriers that continue to constrain it. This includes improving access to land, finance, markets, and infrastructure, while simultaneously recognizing and supporting Indigenous knowledge systems and locally embedded governance structures. Without such interventions, the burden of adaptation will remain disproportionately placed on Indigenous communities, limiting the transformative potential of IE.

Note

1 Summary of interview data from conversation with Indigenous entrepreneurs in South Africa

References

Afrin, T., & Islam, M. S. (2021). Exploring the livelihood pattern of the floating population using the SL framework: A case study of metropolitan Dhaka, Bangladesh. *Journal of the Asia Pacific Economy*, 1–30.

Akbar, N., Abubakar, I. R., & Bouregh, A. S. (2020). Fostering urban sustainability through the ecological wisdom of traditional settlements. *Sustainability, 12*(23), 1–19.

Alcoff, L. (1991). The problem of speaking for others. *Cultural Critique, 20*, 5–32.

Amghani, M. S., Sabouri, M. S., Baghernejad, J., & Norozi, A. (2025). Factors affecting the livelihood sustainability of smallholder farmers in Iran. *Environmental and Sustainability Indicators, 26*, Article 100601.

Carney, D., Drinkwater, M., Rusinow, T., Neefjes, K., Wanmali, S., & Singh, N. (1999). *Livelihoods approaches compared.* Department for International Development.

Chambers, R., & Conway, G. R. (1991). *Sustainable rural livelihoods: Practical concepts for the 21st century* (IDS Discussion Paper No. 296). Institute of Development Studies.

Cheteni, P., & Umejesi, I. (2023). Application of Indigenous knowledge and sustainable agriculture by small-scale farmer households and entrepreneurs in South Africa. *International Journal of Development and Sustainability, 12*(1), 1–18.

Cheteni, P., & Umejesi, I. (2024). Indigenous knowledge in entrepreneurship and cultural tourism in rural areas. *Comparative Sociology, 23*(3), 340–359.

Colbourne, R. (2017). Indigenous entrepreneurship and hybrid ventures. In A. C. Corbett & J. A. Katz (Eds.), *Perspectives and approaches to blended value entrepreneurship: Advances in entrepreneurship, firm emergence and growth* (Vol. *19*, pp. 93–149). Emerald Publishing.

Colbourne, R., Peredo, A. M., & Henriques, I. (2024). Indigenous entrepreneurship? Setting the record straight. *Business History, 66*(2), 455–477.

Commonwealth Network. (2024). *Ezinqoleni.* https://www.commonwealthofnations.org/info/provinces-in-south-africa/kwazulu-natal/ethekwini-metropolitan-area/ugu-district/ezinqoleni/

Cornhill, G., Grussendorff, S., Inglis, R., Kromberg, H., Krone, B., Makkink, P., Peden, M., Prigge, T., & Salomon, M. (2007). *The Maloti Drakensberg experience: Exploring the Maloti Drakensberg route* (1st ed.). Jive Media.

Creswell, J. W., & Creswell, J. D. (2005). *Mixed methods research: Developments, debates, and dilemmas.* Berrett-Koehler.

Creswell, J. W., & Creswell, J. D. (2018). *Research design: Qualitative, quantitative, and mixed methods approaches* (5th ed.). SAGE.

Dana, L.-P. (2015). Indigenous entrepreneurship: An emerging field of research. *International Journal of Business and Globalisation, 14*(2), 158–169.

Deka, N., & Goswami, K. (2025). Indigenous entrepreneurship and agricultural development: Insights from the case of "Darjeeling tea" in Eastern Himalayas. *Climate and Development, 17*(10), 905–920.

Fischer, C., Trono, R. B., & Salinas, R. A. F. (2021). Addressing gaps in socio-economic and natural assets to halt marine turtle egg poaching: A livelihood feasibility study in the Turtle Islands Wildlife Sanctuary, Philippines. *Global Ecology and Conservation,* 32, Article e01913.

Francis, M. (2007). *Explorations in ethnicity and social change among Zulu-speaking San descendants of the Drakensberg Mountains, KwaZulu-Natal* (Doctoral dissertation, University of KwaZulu-Natal).

Francis, M. (2009). Silencing the past: Historical and archaeological colonisation of the Southern San in KwaZulu-Natal, South Africa. *Anthropology Southern Africa, 32*(3–4), 106–116.

Goh, A. H. X. (2012). *A literature review of the gender-differentiated impacts of climate change on women's and men's assets and well-being in developing countries.*

Haidar, M. (2009). *Sustainable livelihood approaches: The framework, lessons learnt from practice and policy recommendations.* Economic and Social Commission for Western Asia (ESCWA) & United Nations Development Programme.

Jurado, T., & Mika, J. P. (2023). Indigenous entrepreneurs' export practices in Aotearoa New Zealand. *Small Enterprise Research, 30*(1), 1–23.

Leal Filho, W., Barbir, J., Gwenzi, J., Ayal, D., Simpson, N. P., Adeleke, L., Tilahun, B., Chirisa, I., Gbedemah, S. F., Nzengya, D. M., Sharifi, A., Theodory, T., & Yaffa, S. (2022). The role of Indigenous knowledge in climate change adaptation in Africa. *Environmental Science & Policy, 136*, 250–260.

Mabizela, S. E. (2022). *Bergville residents' construction of the cultural practice of ukuthwala: A discourse analysis* (Doctoral dissertation, University of South Africa).

Mahama, T. A., & Maharjan, K. L. (2019). Determining the nature and spatial-temporal changes of the livelihood asset pentagon and its relationship with livelihood opportunities in Ghana. *Community Development, 50*(4), 460–483.

Mazzocchi, F. (2020). A deeper meaning of sustainability: Insights from Indigenous knowledge. *The Anthropocene Review, 7*(1), 77–93.

Mersha, A. A., & Van Laerhoven, F. (2016). A gender approach to understanding the differentiated impact of barriers to adaptation: Responses to climate change in rural Ethiopia. *Regional Environmental Change, 16*(6), 1701–1713.

Mohamed Shaffril, H. A., Filho, W. L., Barbosa, L. G., & Burhan, N. A. S. (2025). A systematic literature review on adaptation strategies towards climate change impacts among Indigenous fishers. *Inland Waters,* 1–18.

Natarajan, N., Newsham, A., Rigg, J., & Suhardiman, D. (2022). A sustainable livelihoods framework for the 21st century. *World Development, 155*, Article 105898.

Okhahlamba Local Municipality. (2022). *Okhahlamba Integrated Development Plan (IDP)2022/2023–2026/2027.*

Padilla-Meléndez, A., Plaza-Angulo, J. J., Del-Águila-Obra, A. R., & Ciruela-Lorenzo, A. M. (2022). Indigenous entrepreneurship: Current issues and future lines. *Entrepreneurship & Regional Development, 34*(1–2), 6–31.

Peredo, A. M., & McLean, M. (2013). Indigenous development and the cultural captivity of entrepreneurship. *Business & Society, 52*(4), 592–620.

Peredo, A. M., Anderson, R. B., Galbraith, C. S., Honig, B., & Dana, L. P. (2004). Towards a theory of Indigenous entrepreneurship. *International Journal of Entrepreneurship and Small Business, 1*(1–2), 1–20.

Ratten, V., & Dana, L. P. (2017). Gendered perspective of Indigenous entrepreneurship. *Small Enterprise Research, 24*(1), 62–72.

Scoones, I. (1998). *Sustainable rural livelihoods: A framework for analysis* (IDS Working Paper 72). Institute of Development Studies.

Selelo, M. E. (2023). Indigenous entrepreneurship and the Fourth Industrial Revolution in South Africa. *African Journal of Development Studies, 2023* (SI1), 129.

Senanayake, S. G. J. N. (2006). Indigenous knowledge as a key to sustainable development. *The Journal of Agricultural Sciences, 2*(1), 87–94.

Simiyu, R. R. (2012). *"I don't tell my husband about vegetable sales": Gender dynamics in urban agriculture in Eldoret, Kenya.* African Studies Centre.

Soma, H., Sukhwani, V., & Shaw, R. (2022). An approach to determining the linkage between livelihood assets and housing conditions in urban slums of Dhaka. *Journal of Urban Management, 11*(1), 23–36.

Tengeh, R. K., Ojugbele, H. O., & Ogunlela, O. G. (2022). Towards a theory of Indigenous entrepreneurship: A classic? *International Journal of Entrepreneurship and Small Business, 45*(1), 1–15.

Trang, N. T., & Loc, H. H. (2021). Livelihood sustainability of rural households in adapting to environmental changes: An empirical analysis of the ecological shrimp aquaculture model in the Vietnamese Mekong Delta. *Environmental Development, 39*, Article 100653.

Tuhiwai Smith, L. (2012). *Decolonising methodologies: Research and Indigenous peoples* (2nd ed.). Zed Books.

Wessels, M. (2012). A Bushman voice from the Drakensberg: Zanele Mkhwanazi's story. *Current Writing: Text and Reception in Southern Africa, 24*(1), 58–71.

Yanga, L. P., Mandende, I. P., & Cekiso, M. (2024). A linguistic analysis of spelling errors of selected Indigenous town and village names in the Eastern Cape, South Africa. *Nomina Africana: Journal of African Onomastics, 38*(1), 1–2.

Chapter 9

THE IMPERATIVE OF DECOLONIZING INDIGENOUS ENTREPRENEURSHIP AND POLICY IN AFRICA

Introduction

Decolonization has gained much traction in the epistemic and pedagogical space due to the awareness that there is a need to shed colonial and Eurocentric structures that continue to dominate most of Africa. The necessity to do away with colonial structures has been echoed by Global South thinkers such as Walter Mignolo, Anibal Quijano, Sabelo Ndlovu-Gatsheni, and Sylvia Tamale, to name a few. They recognize the damage that colonial structures have done and continue to do to Africa and the Global South more broadly. These structures often reproduce colonial violence in the form of silencing Indigenous knowledge, Western bias in understanding Africa, and the application of unsuitable policies to drive development (Olaitan and Oloruntoba 2023). The silencing of indigenous knowledge systems yields much damage to the sustainable development of Africa due to the loss of potential recipes for growth embedded in it. Part of this damage is the inability of Africa to harness the gains of entrepreneurship for its socioeconomic development due to the limitations of current entrepreneurship policies. Considering that the African entrepreneurship ecosystem is largely indigenous and based on indigenous practices, it is interesting that entrepreneurship policy in Africa excludes Indigenous voices and Indigenous knowledge. The reason for this is the preference of government and policymakers for Western templates of entrepreneurship policy or the adoption of conventional approaches to entrepreneurship. Both of these tend to ignore the realities of most Indigenous entrepreneurs in Africa, leading to incompatibility between the policy and the intended outcome. Recalling that Western frameworks often reproduce colonial violence and inhibit Indigenous frameworks, policies developed from them are unable to account for the African context. This failure to account for context hinders the participation of Indigenous voices in the creation and ownership of entrepreneurship policy and the acknowledgement of Indigenous knowledge systems.

Based on the foregoing assertion, this chapter examines the subject of entrepreneurship policy (EP) in Africa. It adopts a decolonial lens in its engagement of current entrepreneurship policy in Africa in a bid to advocate for a decolonial entrepreneurship policy that centers Indigenous knowledge and entrepreneurship. The chapter starts by providing an overview of the silencing of indigenous knowledge systems and how this produces epistemicide and an imposition of Western frameworks. It provides this overview to ground its adoption of a decolonial lens as its framework for rethinking entrepreneurship policy in Africa. The next section provides an overview of entrepreneurship policy broadly, while also localizing the discourse on entrepreneurship policy in Africa with examples. This section presents some of the approaches and frameworks for designing EP. The third section examines the limitations and issues with the current discourse on and approach to entrepreneurship policy. The fourth section critically engages the need for an inclusive entrepreneurship policy that fosters the inclusion of Indigenous voices. The fifth section reflects on what a decolonial policy looks like, specifically the advocacy for the participation of Indigenous voices in the creation and ownership. It proposes a decolonial approach to entrepreneurship policy that is grounded in ownership and co-design of the policies by Indigenous communities in Africa. The final section concludes the chapter while proffering relevant recommendations.

Setting the Scene: Epistemicide, Silencing of Indigenous Knowledge, and Decoloniality

We cannot begin a discussion of entrepreneurship policy without acknowledging the erasure of indigenous entrepreneurship and indigenous knowledge within the space. An extensive look into the Eurocentric and Western knowledge framework reveals a centuries-long history of slavery, colonialism, and neo-colonialism, all of which have contributed to the total silencing, or what others might call epistemicide, of indigenous knowledges (Mignolo 2011; Ndlovu-Gatsheni 2018). In order to solidify the knowledge hierarchies between the Global North and the Global South, Kelley (2000:27) points out that colonial domination necessitated an entire style of thinking—a language in which everything that is good, civilized, and advanced is defined and assessed in terms of Europe. Therefore, the coloniality of knowledge played a pivotal role in this process, as it oppressed alternative knowledge systems while imposing Eurocentric ways of thinking. Ndlovu-Gatsheni (2018) opines that because knowledge and power are closely related, control over knowledge production is essential to the maintenance of the unequal global power structures that have existed since the dawn of Euro-North American-centric modernity. The greater consequence of this control is the silencing of alternative knowledge systems through epistemic violence, leading to epistemicide. The core of coloniality of knowledge is the perpetuation of epistemic violence during an encounter with the Other. This led to what Boaventura de Sousa Santos termed epistemicide, i.e., the killing of knowledge (Santos 2014). Spivak (1988) notes that epistemic violence includes both the misinterpretation of colonized subjects' worldview and the imposition of Western knowledge systems. By erasing their history, epistemic violence convinces the subaltern that they have nothing to offer the modern world and that their only option is to follow the enlightened colonists, absorb their knowledge, embrace their worldviews, and become second-class citizens on the outskirts of their own world. Fataar and Subreenduth (2016: 107) see epistemicide as a "metaphor for the epistemological marginalization, or evisceration, of African-centered intellectual traditions in formal education. It is caused by the dominant Western paradigm of knowledge generation, production, and consumption that unproblematically circulates within educational discourse and practice across the African continent as relevant, valuable, and best practice." They also highlight that epistemicide is a consequence of the persistent dominant Western framework of knowledge generation, production, and consumption that uncritically permeates African educational discourse and practice as meaningful, pertinent, and best practice (Fataar & Subreenduth, 2016: 107). Silencing other knowledge systems is essential to the dominance of Western knowledge.

Thus, the silencing of indigenous knowledge systems led to an elevation of Eurocentric knowledge as the universal path to knowledge and development, a notion that is still being reproduced in contemporary times through various means. The epistemicide of indigenous knowledge systems affected all aspects of indigenous societies, as their entire worldview was deemed inferior and subordinate. So, discussions on indigenous entrepreneurship, for instance, were deemed redundant and inconsequential to socioeconomic development in Africa. This redundancy of indigenous entrepreneurship led to the imposition of the Western entrepreneurship model and policy guiding this model on indigenous communities in Africa.

In this vein, this chapter uses a decolonial approach and critique to engage the discourse of entrepreneurship policy in Africa. Decoloniality and decolonial theory have been foundational to most critical studies on re-centering indigenous knowledge and practices across the global south and, more broadly, the decolonization of knowledge production. Quijano (1991) states that "it is necessary to extricate oneself from the linkages between rationality/modernity and coloniality, first of all, and definitely from all power that is not constituted by free decisions made by free people." Mignolo (2011) notes that decolonial thinking emerged as a counterresponse to the foundational elements of modernity/coloniality. He describes decoloniality "as the opening and the freedom from the thinking and the forms of living (economies-other, political theories-other), the cleansing of the coloniality of being and of knowledge; the de-linking from the spell of the rhetoric of modernity, from its imperial imaginary articulated in the rhetoric of democracy" (Mignolo 2011:48). Decoloniality frames the thinking outside of the coloniality/modernity box; it critiques the coloniality embedded in the imposition of Western knowledge frameworks to open up the recentering of other knowledge systems. Its relevance for this chapter lies in its ability to draw out the Western-centrism of current entrepreneurship policy while providing a pathway for developing a decolonial entrepreneurship policy in Africa. Importantly, it sets the course for rethinking entrepreneurship policy in Africa to include excluded voices, center indigenous knowledge, and expand the space for designing entrepreneurship policy.

Understanding Entrepreneurship Policy

Entrepreneurship policy is a broad field, so there are different understandings of what it means. Audretsch et al. (2007) define entrepreneurship policies as strategies or courses of action that the government has put in place to encourage and support entrepreneurial decisions and actions. They are the rules and regulations that facilitate the initiation and sustainability of entrepreneurial endeavors. Cassim et al. (2014) assert that it includes initiatives at different governmental levels to promote entrepreneurship. They add that the capacities of the government and partner agencies to drive entrepreneurship policy vary greatly and always span several government departments. Noailles-Simeon (2017) highlights that entrepreneurship policy covers different areas, such as small business policy, innovation policy, development policy, employment and immigration laws, and intrapreneurship policy. Cassim et al. (2014) contend that since almost all government policies have an impact on small businesses, a significant problem with the concept is that it may refer to policies that influence entrepreneurship. Hart (2003:6) makes a distinction between small business policy and entrepreneurship policy, stating that small businesses are not covered by entrepreneurship policy because, even though many small businesses are small, entrepreneurship policies can cover quite large business lines. According to Stevenson and Lundstrom (2007), there is uncertainty over the definition of entrepreneurship and the extent of entrepreneurship policy in comparison to small and medium-sized enterprises (SMEs) policy. Traditional SME policies, as well as the creation and application of entrepreneurship policies, have been among the policy options (Lundstrom and Stevenson 2005; Smallbone et al. 2010). While entrepreneurship policies concentrate on individual entrepreneurs who may be at varying stages of starting a business, SME policies often focus on a variety of interventions to support small and medium businesses that are already in operation (Stevenson and Lundström 2002; Smallbone et al. 2010). Innovation policy has recently been incorporated into entrepreneurship policy (Stam 2008). Akinyemi and Adejumo (2018) note that while some regulations directly impact entrepreneurs, others are tailored to certain enterprises.

Audretsch and Beckmann (2007) explain that the foundation of entrepreneurship policies is the idea that encouraging entrepreneurship by the public sector boosts employment, boosts economic growth, and improves global competitiveness. This is in consideration that entrepreneurship is hindered by behavioral, institutional, and market challenges. The challenges include administrative expenses related to business registration, information gaps in financial markets that result in liabilities of newness and smallness, and impediments to entry in markets with substantial incumbents (Stinchcombe 1965). The fundamental justification for entrepreneurship policy is that to ensure the economic and social advantages of entrepreneurship, these kinds of obstacles must be addressed.

Akinyemi and Adejumo (2018) opine that entrepreneurship policy includes a local dimension and a national/international dimension; the local policies focus on the sociological side of entrepreneurship, while national or international policies are on the legal side. In general, it includes a framework ethic that permits people to become entrepreneurs and a social component of risk acceptance (Akinyemi and Adejumo 2018). Entrepreneurship policies are purely practical and efficiency-driven; they are not theoretical methods (Noailles-Simeon 2017). Many countries have put in place both general and targeted policies to encourage entrepreneurial activity in response to the need for entrepreneurship policies. In many economies, general policies like tax rates, labour legislation, and market restrictions have greatly influenced the atmosphere for entrepreneurship. Akinyemi and Adejumo (2018) state that certain policies have been specifically designed to encourage entrepreneurship. Additionally, studies have demonstrated that public policies concerning taxation and company rules frequently have an impact on entrepreneurial endeavors (Acs and Szerb 2007). Entrepreneurial activity is impacted by monetary policy, tariffs, and taxes. For example, the government sells Treasury bills to the general public when it decides to take money out of the economy. According to Akinyemi and Adejumo (2018), this always results in a decrease in the amount of money in circulation, influences investors' willingness to disburse cash, and eventually cripples entrepreneurial activities.

The goal of entrepreneurship policy should not only be increasing start-up rates, but the caliber of newly established enterprises should also be improved. Resources should be directed towards start-ups with the potential to expand and sustain themselves if policies aim to achieve greater economic and social benefits. But policy should not just target the "gazelles" or the most active start-ups. These start-ups are rare, yet they have a disproportionate impact on the diffusion of innovation and the creation of jobs. Therefore, keeping regional differences in mind while designing entrepreneurship strategies will often be beneficial. This includes business support programs (such as start-up grants, training programs, and export support), which frequently exhibit the highest take-up rates in the most entrepreneurial regions, as well as systemic interventions (such as fiscal and economic policies), which may have varying effects on various business community segments (OECD 2020). When designing, selecting, and implementing policies, entrepreneurship should be a clear priority. In addition to affecting both high-tech and low-tech economic activity, entrepreneurship policy covers activities at all levels of government, from local to national, and includes governance capabilities such as poverty alleviation, economic development partnerships, and regulatory policies (Hart 2003). Policymakers typically want to increase the degree of entrepreneurship to promote a socially optimal level of such ventures (Reynolds et al., 2000). According to Hart (2003), it frequently seeks to accelerate improved governance, for example, by creating networks of possible clients and service providers, whose existence lessens the uncertainty faced by business owners. He issued a warning, saying that not all governmental policies that influence the environment and availability of prospective business owners are entrepreneurship policies. The degree to which governments decide to offer direct aid, in contrast to programs that reduce the barriers to entrepreneurship, is another important consideration in policy decisions (Cassim et al., 2014).

A brief overview of the evolution of EP traces the earliest mention of entrepreneurship policy to the late 1800s. Gilbert et al. (2004:313) note that during this period, EPs were focused on "harnessing the market power of large corporations through a triad of policy instruments including regulation, antitrust, and government ownership." Following this period, there was the New Deal, which was formulated in the 1930s and was one of the earliest examples of entrepreneurship policy, with more policies emerging after this period (Noailes-Simeon 2017). Gilbert et al. (2004:313) argue that the 1980s ushered in the downsizing or closing of public agencies necessitated by the privatization of publicly owned enterprises. This period saw the rise of privately owned enterprises, thereby influencing the nature of EPs that they formulated. In accordance with this shift, the 1990s saw a growing interest in entrepreneurship policy research. Hart (2003) states that entrepreneurship policy emerged as a "politically correct" policy topic during this period. From the early 2000s to date, EPs have focused on technology-driven enterprises and small, medium, and micro enterprises. Gilbert et al. (2004) contend that there was a shift in the nature of entrepreneurship policies between the early 1900s and early 2000s; EP in the former worked to protect big businesses, while during the latter, EP focused on enabling the growth of new and small businesses.

Differently stated, entrepreneurship policy has gone through several shifts over the years (OECD 2020). Early EPs started with protecting large corporations, with a progression into enabling the growth of small businesses, and more recently, cutting-edge focus on comprehensive packages for start-ups with growth potential. The significance of entrepreneurship policy as a policy area is encapsulated by the presence of small and medium-sized business (SME) enterprises, development banks, and economic development agencies at the national, regional, and local levels, as well as government ministries of economy, industry, and business development. Entrepreneurship policy is a crucial component since there are undoubtedly examples where public policy and governance were the primary factors influencing the quantity and caliber of entrepreneurship (Hart 2003).

Entrepreneurship Policies in Africa

In Africa, there have been different entrepreneurship policies that have been implemented to encourage and strengthen entrepreneurship. The African Union (AU), under the New Partnership for Africa's Development (NEPAD) and the African Union's strategy for MSMEs, is promoting policies, initiatives, and programs that encourage the growth of an enabling entrepreneurial ecosystem. The 2017 Africa Entrepreneurship Policy Forum recognized the importance of effective implementation of entrepreneurship policies to the growth of the entrepreneurial ecosystem (UNCTAD 2017). At the national level, African governments are implementing specific measures that aim to remove regulatory barriers, improve education and skill development, facilitate access to financing, foster innovation and technology exchanges, and emphasize public awareness, local support, and regional networks. For instance, in South Africa, the government passed the 1996 Small Business Act (Ladzani and Van Vuuren 2002) in acknowledgement of the significance of entrepreneurship to economic development. The 2009 Global Entrepreneurship Monitor (GEM) reported that one of the main obstacles to the country's entrepreneurial activity is education in general and entrepreneurship education in particular. In acknowledging the significance of entrepreneurship education for entrepreneurship, the government introduced entrepreneurship education in the country (Cassim et al 2014). In Mauritius, the Small and Medium-Size Enterprise Master Plan was introduced to boost entrepreneurship in the country. The main objectives of the Master Plan are to improve SME competitiveness and growth, foster high growth potential, upgrade skills and job opportunities by addressing skill mismatch and upgrading human capital, improve design and value addition by supporting research and developing innovation, and increase market access and exports (UNCTAD 2017). In Rwanda, the recognition that an integrated approach and a targeted, cogent entrepreneurship policy are required to foster an environment that supports the growth of innovation and entrepreneurship led to the development of a new EP in 2020. The new entrepreneurship policy aims to assist private sector entrepreneurs and provide the conditions for private sector dynamism and innovation—both of which are essential for a developing economy (Rwanda MTI 2020). In Nigeria, due to its large entrepreneurship sector, different EPs have been implemented, such as the Government Enterprise and Empowerment Program (GEEP), N-Power, the Small and Medium Enterprises Development Agency (SMEDAN), and the You-WIN program. These policies were implemented to encourage entrepreneurship by making it easier for SMEs to obtain capital and other resources they require (Oliyide 2012). These examples demonstrate the existence of EPs in Africa and the different levels of policymaking.

Approaches to entrepreneurship policy

Different actors and stakeholders have dabbled in the field of developing an appropriate framework or approach to guide the development of entrepreneurship policy. For example, Stevenson and Lundström (2005:60) created one of the earliest entrepreneurship policy frameworks to guide the development of EP. These included promotion of entrepreneurship: encouraging a culture of entrepreneurship, education of entrepreneurship: encouraging entrepreneurship in schools, startup environment: lowering entrance hurdles, removing business and tax reform impediments, startup and seed money financing: lending money to start-up companies, and seed money and company support initiatives for startups: offering incubators, networks, and start-up portals, as well as other company start-up assistance. In 2009, the United Nations Conference for Trade and Development (UNCTAD) investigated the creation of frameworks for entrepreneurship policy, resulting in the development of a framework that was applicable to all countries. This framework included research and development, technology transfer, awareness and network building, access to finance, entrepreneurial education and skills, general entrepreneurship policy (including enabling environment), and regulatory environment.

In 2017, Brown and Mason (2017) established the four essential elements of EP that are important to develop the potential of entrepreneurs in the entrepreneurial ecosystems. These included provision of incubation, acceleration, coaching, and mentoring services to entrepreneurs; provision of financial resources (banks, business angels), knowledge, and opportunities for collaboration (large firms, research institutions); building networks and connections within the ecosystem (professional associations, business brokers); and imbibing an entrepreneurial orientation, which includes a culture that encourages entrepreneurship. OECD (2020) provides a comprehensive approach for the development of EP by delineating them into three modes. First, policies that target institutional conditions, second, policies that offer direct assistance to entrepreneurs and start-ups, and third, policies that seek to enhance entrepreneurial ecosystems. These approaches explain the different levels of interventions for entrepreneurship policy. For example, policies that target institutional conditions aim at reducing administrative and regulatory obstacles to business and fostering positive attitudes and motives among prospective entrepreneurs are the main goals of policies intended to enhance institutional conditions for entrepreneurship. Initiatives to foster an entrepreneurial culture, enhance the tax code to encourage entrepreneurship and investment in start-up companies, strengthen competitive conditions, and guarantee that the regulatory environment supports the establishment and expansion of businesses by a variety of entrepreneurs are all included in this category of policy intervention. A national focus is common in policies aimed at improving institutional conditions.

Policies that offer direct assistance to entrepreneurs entail programs aimed specifically at entrepreneurs and constitute the second category of policies for productive entrepreneurship. They seek to solve the main obstacles to starting a business by assisting entrepreneurs in obtaining the resources (knowledge, capital, networks, etc.) needed to launch and expand their enterprises. Programs that offer direct assistance to entrepreneurs can be general, available to all aspiring business owners, or they can be customized to meet the needs of particular demographic groups that are underrepresented among entrepreneurs or face particular barriers to starting their businesses (e.g., women, youth, the unemployed, and people living in rural regions). Programs may combine many forms of support and employ a variety of interventions to assist entrepreneurs. Policies that seek to enhance entrepreneurial ecosystems encourage the growth of entrepreneurial ecosystems is to create an extensive and integrated environment that is conducive to the launch and expansion of businesses. Targeting a broad spectrum of players, these policies frequently include the coordination of several lines of action and directly or indirectly assist other kinds of interventions covered above (e.g., initiatives for institutional stability and in helping entrepreneurs).

Issues with Conventional Entrepreneurship Policy in Africa

Having explained entrepreneurship policy, contextualized EPs in Africa, and presented the different approaches/frameworks, it is necessary to highlight the shortcomings of conventional approaches for promoting Indigenous entrepreneurship policy in Africa. Fonrouge (2022) argues that there is a "Western bias" in entrepreneurship policy when the reductive Western perspective is applied to examine what is perceived as a widely shared kind of business. The repercussion of this Western bias is the inability of these policies to produce the intended outcome. There is also the absence of contextual specificity in these policies due to the mismatch between the approach/framework used for the development and the ecosystem they are developed for (Banerjee 2021). Germain and Jacquemin (2017) argue that the shortcomings of conventional entrepreneurship policies are widely acknowledged due to their insufficiency to contextualize entrepreneurship activities stemming from their embeddedness in Western knowledge structures.

Smallbone (2016) details the lack of context specificity in conventional approaches to entrepreneurship policy. He notes that a recent trend in the literature on entrepreneurship has been the demand for a clearer focus on context, which has, in the case of some theoretical methods, been somewhat overlooked. The historical, social, institutional, and economic environments all interact to form a complex and singular whole, which is referred to as context (Welter 2011). Different countries, regions, and contexts face different demands, strengths, weaknesses, dangers, and opportunities; therefore, the context in which entrepreneurship can have a positive impact on the development of these places also differs. More precisely, the state of entrepreneurship today and its trends are part of the framework for entrepreneurship policy. The country's needs may be impacted to encourage more entrepreneurship. Additionally, the features of these regions have an effect, which could affect how simple or complex the policies should be.

Furthermore, context is crucial when considering the process of formulation and implementation of entrepreneurship policy. There has been a lot of literature written about this exercise, which reflects its potentially contentious character. Without compatibility between the institutional frameworks of the policy and the intended demography, the latter strategy is unlikely to be successful. This is true in terms of how these institutions act, as well as in terms of their formal institutional framework (Smallbone 2016: 207). In discussions on the relevance of Indigenous institutions' compatibility for entrepreneurship policy, the question of whether these Indigenous institutions will be adequate to meet the demands of the current entrepreneurial ecosystem is raised. Most literature on entrepreneurship in Africa tends to focus on and parrot the notion that the entrepreneurial ecosystem in Africa is informal (Olaitan 2024). This bias of informality affects the kind of approaches that are adopted for developing policies on entrepreneurship, how entrepreneurship in Africa is understood, and suggestions for growing the African entrepreneurial space. This bias also feeds the misconception that most entrepreneurs in Africa lack the requisite entrepreneurial education. According to GEM (2005), there is typically a negative correlation between education and entrepreneurial activity in low-income countries and a positive correlation in high-income ones. People with less education launch businesses in low-income countries, while those with more education launch businesses in high-income countries. This implies that although fewer educated entrepreneurs are engaging out of necessity, more educated entrepreneurs are pursuing opportunity-based companies. Dhewa (2022) contends that one of the colonial legacies that African countries face is the association of formal education with entrepreneurship. She argues further that several African governments have begun establishing university-based innovation and entrepreneurship institutes. Despite being a good cause, this has created the idea that entrepreneurship is a subject best taught in academic institutions. The degree of entrepreneurship in Africa, on the other hand, indicates that entrepreneurship has little to do with formal education or a university degree (Dhewa 2022).

Apart from Western bias and lack of context specificity, other factors affect the effectiveness of entrepreneurship policy. Oosthuizen (2010) notes that numerous programs have been launched by the public and private sectors to encourage the growth and expansion of entrepreneurship. Even while the value of entrepreneurship to economic growth is becoming more widely acknowledged, these measures have not produced the expected outcomes. Endeavour SA (2010:5) states that South African cultural norms do not encourage entrepreneurs who have failed. It is suggested that this stereotype hinders the development of an entrepreneurial attitude among the populace. Furthermore, the GEM (2009) report noted that one of the main obstacles to small business expansion in South Africa is the country's stringent labour laws. In South Africa, government regulations do not align with initiatives to encourage entrepreneurship. According to the GEM Report (2009), insufficient efforts are being made to lessen the regulatory environment's negative impacts. Women continue to engage in entrepreneurship in low-productivity industries. Furthermore, the biggest obstacles for women are determined to be inequities and unfavorable social and cultural contexts (UNCTAD 2017). There are differing opinions about the causes of these results, and it is evident that entrepreneurship policy has come under criticism for being ineffective. Arshed et al. (2014) focus on the process of formulating policy as one of the factors that might affect its effectiveness.

Mapping an Inclusive Entrepreneurship Policy in Africa

In light of the failures and limitations of conventional entrepreneurship policy, it is important to delineate an innovative and grounded means of formulating entrepreneurship policy that encourages context-specificity and inclusivity in Africa. OECD (2020) advocates for the inclusion of a contextual dimension in entrepreneurship policies to address regional differences in the type and severity of impediments to entrepreneurship, as well as to account for structural differences among regions. IOE (2005) posits that when creating policies to promote entrepreneurship, two fundamental tenets must be considered. First, policies need to be modified to accommodate various national contexts. The main backdrop for all business activity in certain developing countries will be the blatant absence of market prospects, and governments must adapt to this fact. Second, policy must consider the diversity of the entrepreneurial ecosystem because entrepreneurs are not a monolithic group. Increased access to seed capital will be crucial for certain groups of entrepreneurs, while for more experienced entrepreneurs, the most crucial policy may be to open up opportunities for public-private investments or modify government procurement processes; at the other extreme, it may simply mean making it simpler to register a business.

Cassim et al. (2014) recognize that African countries need a policy framework that integrates innovation for global competitiveness with local economic growth because their unique conditions and policy objectives necessitate a horizontally rooted entrepreneurship policy. Effective policies avoid a top-down structure and instead embrace a comprehensive approach, actively and collaboratively engaging stakeholders (Autio 2016). For instance, they should promote forums that connect entrepreneurs with other players (e.g., pitching events, training, and matchmaking) to facilitate the development of networks within the ecosystem (Isenberg, 2011). The demands of businesses at various phases of their development should be considered when designing interventions. Moreover, policy must typically concentrate at the sub-national level due to the significance of ecosystems' geographical dimension. To make it easier to coordinate entrepreneurial ecosystems, there should be specialized independent coordinating bodies available (OECD 2020). Policies intended to boost and scale up entrepreneurship have immediate and direct effects on governments in terms of broadening the tax base. In that regard, the advantages are evident and palpable. However, the broader advantages, such as job creation, reducing social pressures, giving other businesses opportunities to be suppliers, more choice, and greater economic integration that promotes trade and investment relationships, are just as significant (and can also result in increased regional stability).

Ahmed and Nwankwo (2013) explain that an inclusive policy must use enterprise-led initiatives to create a new generation of entrepreneurs who are free from institutionalized and structured governmental barriers and empowered by enhanced support networks that not only acknowledge their unique disabilities but also offer workable solutions instead. Considering the remarkable success of women entrepreneurs' business ventures and the fact that they make up the majority of the population in these countries, focusing on them will significantly improve the use of enterprise solutions in Africa to reduce poverty and maintain sustainable development. Entrepreneurship must be conceptually understood, and the fundamental flaws in national planning policies must be reevaluated to be viewed as a solution to the diverse and complex market and development difficulties of African countries. Importantly, these consist of monetary, fiscal, and educational policies. This strategy must include rural development projects that provide a functioning telecommunications system, accessible roads, water, and a dependable power source.

A Decolonial Entrepreneurship Policy in Africa

Adopting a decolonial approach to override the embedded Western bias and also ensure the re-centering of Indigenous knowledge in entrepreneurship policy is central to achieving an inclusive entrepreneurship policy in Africa. In the broader discussion on entrepreneurship policy, the participation of entrepreneurs and the relevance of Indigenous knowledge systems are largely ignored. Even though the importance of Indigenous knowledge systems has been widely acknowledged in Africa, Odey (2020:264) states that "much more needs to be done to further show that development is no longer the exclusive domain of Western science/global knowledge," which in this case includes creating inclusive policies for entrepreneurship. First, it is crucial to decolonize Africa's knowledge system. Second, it is important to highlight the full potential of the various aspects of indigenous knowledge systems. To comprehend entrepreneurship in Africa, historical and contextual factors must be considered.

Fonrouge (2022) explains that a decolonial method begins with a critique of existing entrepreneurship classifications because these classifications reinforce inequality by maintaining an asymmetry of power and authority rather than allowing for greater freedom. These classifications do not allow for alternative voices or forms of entrepreneurs to be covered in the policy, and the effect of this erasure needs to be considered. Martinez and Jayawarna (2020) add that one of the possible consequences of these classifications is that they occasionally overlook the unseen aspects of entrepreneurship, such as sex workers or domestic workers and women cross-border traders, who are rarely studied despite being independent business owners or self-employed individuals. By emphasizing a uniform view of the entrepreneurial ecosystem, 'other' entrepreneurs are disregarded since they do not fit in and occasionally do not even consider themselves to be entrepreneurs (Verduijn et al., 2017). Fonrouge (2022) argues that by acknowledging that the Global South and the 'rest' are not uniform, just as the North is, a decolonial perspective goes one step further. A decolonial approach to entrepreneurship policy promotes the visibility and inclusion of women and indigenous communities by going beyond Western classifications that pit "classic" and "minoritized" entrepreneurs against one another from an "us" versus "them" standpoint. These asymmetrical terms ultimately serve to reinforce and prolong the "colonized" and "colonizer" opposition and, consequently, the games of domination (Jack et al., 2011). Recent entrepreneurship trends, like the gig economy, show that some entrepreneurs are caught in unequal power dynamics and fall outside of established categories.

To drive a decolonial entrepreneurship policy, it is important to re-center Indigenous and non-Western voices in the discourse and process of its development and implementation. This will entail a detailed understanding of how Indigenous communities perceive entrepreneurship and how ideas derived from Indigenous ethos affect wealth, success, and, more generally, resources (Ul Haq and Westwood, 2012; Gundolf and Jansen, 2021). Incorporating Indigenous knowledge into entrepreneurship policy can help dissolve the distinction between human and non-human resources. By teaching researchers to employ their critical thinking perspective with a 'dual perspective' approach that unites the agendas of Indigenous communities and those of primarily non-Indigenous researchers, a decolonial entrepreneurship policy that challenges the biased and fragmented nature of dominant Western perspectives can be achieved (Asselin and Basile, 2018). Also, embracing indigenous institutions of entrepreneurship will bid well for developing a decolonial entrepreneurship policy. Smallbone (2016) warns that maintaining dependence on Indigenous institutions without adapting them to contemporary structures hinders the effectiveness of these institutions in meeting the needs of Indigenous entrepreneurs at the current time. This is especially true when it comes to small businesses (Ibid 208).

The exclusion of Indigenous voices in discussions on entrepreneurship policy can be traced to the non-recognition of the agency of Indigenous peoples and the relevance of Indigenous knowledge systems. The views and experiences of those most vulnerable must be prioritized in the design and development of these policies. The basis for recognizing the agency of Indigenous peoples as agents of change who can contribute meaningfully to these discussions, and the development of entrepreneurship policy, is predicated on the acknowledgement of Indigenous knowledge systems as a legitimate system of knowing. However, what we have seen is the back-benching of Indigenous peoples across the continent on matters that gravely affect them. Changes that threaten the existence of Indigenous peoples are widespread in the African region, while their lifestyles, cultures, and human rights are generally not given adequate recognition in the political and policy frameworks of their countries.

The critique of the dominance of mainstream Western approaches to entrepreneurship policy enables us to call for a more inclusive and contextual approach to designing entrepreneurship policy in Africa. Without inclusive and contextual nuance, there is a tendency to reproduce conventional policy approaches that do not account for the realities of indigenous entrepreneurs. The co-designing and ownership of entrepreneurship policy ensure that Indigenous entrepreneurs and their approach to entrepreneurship are adequately reflected in the policy process. Additionally, their participation in the process addresses the dominance of conventional approaches by opening the door to Indigenous approaches to entrepreneurship policy, ultimately de-centering conventional approaches. Considering that most of the shortcomings of conventional approaches to entrepreneurship policy feed off their inability to account for the entrepreneurship ecosystem in Africa, research and literature on entrepreneurship policy must engage Indigenous entrepreneurship and Indigenous approaches to entrepreneurship policy. This includes the recognition of the diverse ways indigenous peoples engage in entrepreneurship, trade, and the economy. The lens through which indigenous knowledge and indigenous entrepreneurship are understood is skewed and fundamentally flawed, necessitating a deconstruction and reconstruction. At the center of this deconstruction and reconstruction is the imperative for decolonization, i.e., doing away with colonial models and approaches to usher in indigenous knowledge and models. Decolonizing entrepreneurship policy reduces the reliance of African governments on Western institutions for guidance on policy design and development, as they can tap into the array of Indigenous knowledge systems to enrich this process. While this chapter echoes the need for a decolonial approach to entrepreneurship policy, it does not suggest an outright rejection of all conventional approaches. A complementary effort is best suited to address the current limitations and failures of these conventional approaches to entrepreneurship policy.

Conclusion

The recognition that entrepreneurship is pivotal to development in Africa has motivated African governments to formulate and implement different policies on entrepreneurship. These policies are meant to drive the growth of the entrepreneurial ecosystem by providing a conducive business environment, regulation, partnerships, etc. The idea behind having policies to support the growth of entrepreneurs is not bad in and of itself; however, the approaches used to design and formulate these policies are often not inclusive and contextually nuanced. This is because conventional approaches to the design and development of entrepreneurship policy are not based on localized knowledge and are premised on a skewed understanding of the entrepreneurial ecosystem in Africa. Central is the ignorance of Indigenous entrepreneurship and Indigenous knowledge systems as important contributors to the development of entrepreneurship policy. The effect is the development of policies that do not address the intrinsic needs of entrepreneurs in Africa, while at the same time impeding their growth potential. For this cause, this chapter adopted a decolonial lens to rethink the discourse on entrepreneurship policy in Africa. It started by engaging the silencing of indigenous knowledge on entrepreneurship as a consequence of epistemicide and the imposition of Western frameworks. It provided a conceptual understanding of entrepreneurship policy while contextualizing the discourse on entrepreneurship policy in Africa. The chapter engaged the limitations and failures in the current discourse on entrepreneurship policy and noted that there is a Western bias embedded in current approaches to entrepreneurship policy. This bias often manifests in the notion that entrepreneurship in Africa is mostly informal, as they do not fit within the parameters of formal entrepreneurship.

Consequently, there is a reproduction of this bias in policy, which ultimately hinders the effectiveness of policies in fostering entrepreneurship growth. Based on this, the chapter argued for a decolonial approach to entrepreneurship policy that centers context, the inclusion of indigenous voices, and the application of indigenous knowledge systems. It notes that a decolonial approach to entrepreneurship policy allows for a contextually nuanced, inclusive, and representative policy. Such a policy should be co-designed, created, and responsive to the needs of Indigenous entrepreneurs. Without a decolonized approach to entrepreneurship policy, Africa will struggle to harness the potential of entrepreneurship for sustainable development. In lieu of the discourse of epistemicide and silencing of knowledge, a decolonial lens to entrepreneurship policy fosters a re-centering of indigenous knowledge and indigenous entrepreneurship. The recentering of indigenous voices in EP in Africa has never been more necessary than in the 21st century. Africa is witnessing an expansion of its informal economy due to the increased number of people resorting to entrepreneurship in response to growing unemployment. Therefore, it becomes necessary for governments and relevant stakeholders to adopt a decolonial approach to EP while devising means of substantively including indigenous voices in the agenda. This should come with the assurance that their knowledge system, practices, and worldview will be represented in the policy process. Noting that the inclusion of indigenous voices in entrepreneurship policy opens the space for decolonizing entrepreneurship policy design and development, a decolonial EP therefore consolidates efforts to maximize the gains of entrepreneurship for socioeconomic development in Africa.

References

Acs, Z. J., & Szerb, L. (2007). Entrepreneurship, economic growth, and public policy. *Small Business Economics, 28*(2–3), 109–122.

Ahmad, A., & Nwankwo, S. (2013). Entrepreneurship development in Africa: An overview. *World Journal of Entrepreneurship, Management and Sustainable Development, 9*(2–3), 82–86.

Akinyemi, F., & Adejumo, O. (2018). Government policies and entrepreneurship phases in emerging economies: Nigeria and South Africa. *Journal of Global Entrepreneurship Research, 8*(35), 1–18.

Asselin, H., & Basile, S. (2018). Concrete ways to decolonize research. *ACME: An International Journal for Critical Geographies, 17*(3), 643–650.

Audretsch, D. B., Grilo, I., & Thurik, A. R. (2007). Explaining entrepreneurship and the role of policy: A framework. In D. B. Audretsch, I. Grilo, & A. R. Thurik (Eds.), *Handbook of research on entrepreneurship policy* (pp. 1–17). Edward Elgar Publishing.

Autio, E. (2016). *Entrepreneurship support in Europe: Trends and challenges for EU policy* (Report for EU DG Growth).

Banerjee, S. B. (2021). Decolonizing management theory: A critical perspective. *Journal of Management Studies, 59*(4), 1074–1087.

Brown, R., & Mason, C. (2017). Looking inside the spiky bits: A critical review and conceptualisation of entrepreneurial ecosystems. *Small Business Economics, 49*(1), 11–30.

Cassim, S., Soni, P., & Karodia, A. (2014). Entrepreneurship policy in South Africa. *Arabian Journal of Business and Management Review, 3*(9), 29–45.

Dhewa, C. (2022, December 13). African mass food markets decolonising entrepreneurship. *Zimbabwe Independent.*

Endeavor South Africa. (2010). *State of entrepreneurship in South Africa.* In Gordon Institute of Business Science (Ed.). First National Bank.

Fataar, A., & Subreenduth, S. (2016). The search for ecologies of knowledge in the encounter with African epistemicide in South African education. *South African Journal of Higher Education, 29*(2), 106–121.

Fonrouge, C. (2022). Postcolonial and decolonial approaches in entrepreneurship: Critical avenues for researchers. *Review of Entrepreneurship, 21*(3), 39–46.

Global Entrepreneurship Monitor. (2009). *Tracking entrepreneurship in South Africa: A GEM perspective.* Global Entrepreneurship Monitor.

Germain, O., & Jacquemin, A. (2017). Voies et voix d'approches critiques en entrepreneuriat. *Revue de l'Entrepreneuriat, 16*(1), 7–18.

Gilbert, B. A., Audretsch, D. B., & McDougall-Covin, P. P. (2004). The emergence of entrepreneurship policy. *Small Business Economics, 22*(3), 313–323.

Gundolf, K., & Janssen, F. (2021). *Entrepreneuriat, spiritualité et religions.* De Boeck Supérieur.

Hart, D. M. (2003). Entrepreneurship policy: What it is and where it came from. In D. M. Hart (Ed.), *The emergence of entrepreneurship policy: Governance, start-ups, and growth in the U.S. knowledge economy.* Cambridge University Press.

International Organisation of Employers. (2005). *Approaches and policies to foster entrepreneurship.*

Isenberg, D. (2011). *The entrepreneurship ecosystem strategy as a new paradigm for economic policy: Principles for cultivating entrepreneurship.* The Babson Entrepreneurship Ecosystem Project.

Kelley, R. D. G. (2000). A poetics of anticolonialism. In A. Césaire, *Discourse on colonialism* (pp. 7–28). Monthly Review Press.

Ladzani, W. M., & Van Vuuren, J. J. (2002). Entrepreneurship training for emerging SMEs in South Africa. *Journal of Small Business Management, 40*(2), 154–161.

Lundström, A., and Stevenson, L. (2002). *On the Road to Entrepreneurship Policy.* Volume 1 of the Entrepreneurship Policy for the Future Series. FSF.

Lundström, A., & Stevenson, L. (2005). *Entrepreneurship policy: Theory and practice.* Springer.

Martinez Dy, A., & Jayawarna, D. (2020). Bios, mythoi, and women entrepreneurs: A Wynterian analysis of the intersectional impacts of the COVID-19 pandemic on self-employed women and women-owned businesses. *International Small Business Journal, 38*(5), 391–403.

Mignolo, W. D. (2011). Epistemic disobedience and the decolonial option: A manifesto. *Transmodernity, 1*(2), 44–66.

Ndlovu-Gatsheni, S. J. (2018). Seek ye epistemic freedom first. In S. J. Ndlovu-Gatsheni (Ed.), *Epistemic freedom in Africa: Deprovincialization and decolonization.* Routledge.

Nkomo, S. M. (2011). A postcolonial and anti-colonial reading of "African" leadership and management in organization studies: Tensions, contradictions, and possibilities. *Organization, 18*(3), 365–386.

Noailles-Siméon, P. (2017). Entrepreneurship policy. In E. G. Carayannis (Ed.), *Encyclopedia of creativity, innovation, and entrepreneurship.* Springer.

Odey, M. (2020). Understanding Igede Indigenous knowledge systems and the future of Igede cultural heritage in Benue State, Nigeria. In S. O. Oloruntoba, A. Afolayan, & O. Yacob-Haliso (Eds.), *Indigenous knowledge systems and development in Africa.* Palgrave Macmillan.

OECD. (2020). *International compendium of entrepreneurship policies.* OECD Publishing.

Olaitan, Z. M. (2024). *The relevance of African Indigenous knowledge systems for development in Africa.* AIKRN Policy Brief Series.

Olaitan, Z. M., & Oloruntoba, S. O. (2023). Beyond reaction: (Re)-imagining African agency in the decolonization of knowledge. In N. Andrews & N. Khalema (Eds.), *Decolonizing African pedagogies: Knowledge production, epistemic imperialism, and Black agency* (pp. 31–53). Springer.

Oliyide, O. (2012). Law, credit risk management, and bank lending to SMEs in Nigeria. *Commonwealth Law Bulletin, 38*(4), 673–695.

Oosthuizen, C. (2010). *Where are SA's entrepreneurial leaders?*

Quijano, A. (1991). Colonialidad, modernidad/racialidad. *Perú Indígena, 13*(29), 11–29.

Republic of Rwanda Ministry of Trade and Industry. (2020). *Entrepreneurship development policy: Developing an effective entrepreneurship and MSME ecosystem in Rwanda.*

Republic of South Africa. (1996). *National Small Business Act No. 102 of 1996.* Government Gazette.

Santos, B. de S. (2014). *Epistemologies of the South: Justice against epistemicide.* Routledge.

Smallbone, D. (2016). Entrepreneurship policy: Issues and challenges. *Small Enterprise Research, 23*(3), 201–218.

Smallbone, D., Kitching, J., & Athayde, R. (2010). Ethnic diversity, entrepreneurship and competitiveness in a global city. *International Small Business Journal*, 28(2), 174–190.

Spivak, G. C. (1988). Can the subaltern speak? In C. Nelson & L. Grossberg (Eds.), *Marxism and the interpretation of culture* (pp. 271–313). Illinois University Press.

Stam, E. (2018). Measuring entrepreneurial ecosystems. In A. O'Connor, E. Stam, F. Sussan, & D. B. Audretsch (Eds.), *Entrepreneurial ecosystems: Place-based transformations and transitions* (pp. 173–197). Springer.

Stevenson, L., & Lundström, A. (2005). *Entrepreneurship policy: Theory and practice.* Springer.

Stevenson, L., & Lundström, A. (2007). Dressing the emperor: The fabric of entrepreneurship policy. In D. B. Audretsch, I. Grilo, & A. R. Thurik (Eds.), *Handbook of research on entrepreneurship policy* (pp. 94–129). Edward Elgar Publishing.

Stinchcombe, A. L. (1965). Social structure and organizations. In J. G. March (Ed.), *Handbook of organizations* (pp. 142–193). Rand McNally.

Ul-Haq, S., & Westwood, R. (2012). The politics of knowledge, epistemological occlusion, and Islamic management and organization knowledge. *Organization, 19*(2), 229–257.

United Nations Conference on Trade and Development. (2009). *Key components of entrepreneurship and innovation policy frameworks.*

United Nations Conference on Trade and Development. (2017). *Africa entrepreneurship policy forum: Entrepreneurship policies for regional inclusive and sustainable development.*

Verduyn, K., Dey, P., & Tedmanson, D. (2017). A critical understanding of entrepreneurship. *Revue de l'Entrepreneuriat, 16*(1), 37–45.

Chapter 10

LEVERAGING INDIGENOUS KNOWLEDGE SYSTEMS TOWARD A NEW SOCIOECONOMIC DEVELOPMENT AGENDA IN AFRICA

Introduction

As African countries continue to confront the crisis of development, questions arise as to why the various endogenous and exogenous policy prescriptions adopted since independence have not led to any significant improvement in the lives of Africans. Although trillions of dollars have been generated through rent from the extraction of raw materials as well as foreign aid, Africa remains the poorest region of the world, with more than half of its population of 1.3 billion still living in poverty (World Bank, 2024; Moyo, 2009, Collier 2007; Easterly, 2007). To be sure, the adoption of conventional of policies framed under neoclassical economic theory and its variants under neoliberalism has fostered economic growth in terms of Gross Domestic Product, the various growth spurts have at best benefited members of the transnational capitalist class made up of ruling elites, representatives of the multinational companies and petit bourgeoisies who have access to those who hold the lever of power,(Fioramonti, 2013, Oloruntoba 2016, Robinson 2004). For instance, from 2004 to 2014, several African countries recorded a high rate of GDP growth that the leading world media and institutions celebrated as a miracle (Economist, 2011; IMF, 2011). As we note below, these growths were not reflective of any fundamental shift in the structure of Africa's economy but a reproduction of the old pattern of commodity exports, with a tinge of changes in the services sector.

The premature celebration of the growth miracle has been criticized for its lack of penetration and limited structural transformation (Taylor, 2014). As Taylor argues, the growth was mostly fueled by commodity exports and depended mostly on overaccumulation in the emerging economies of Brazil, Russia, India, China, and, later, South Africa. Scholars have examined the crisis of development in Africa from various perspectives. The earlier and dominant strands of literature have focused on the state and how its rentier nature fosters corruption (Joseph, 2015; Adebanwi, 2012; Adebanwi and Obadare, 2010, Bayart 2009). Others in the dependency school adopted a structuralist view and showed how exogenous and historical factors such as slavery, imperialism, and colonialism are responsible for maldevelopment, underdevelopment, and dependent development in Africa (Amin, 1990; Ake, 1981; Rodney, 2018).

International organizations like the World Bank and the International Monetary Fund on the other hand posit that lack of socioeconomic development in post independent Africa owed its origin to the political economy structure in which the state played active role in establishing state led corporations, engaged in import substitution industrialization, invested massively in social policies such as free access to education and health, among others (see World Bank Berg Report 1981). The (mis) diagnosis of African economic problems by these institutions and the need for loans by African governments created conditions that tied the state to the adoption of neoliberal policy prescriptions (if not) directions under which many of these countries have been trapped since the 1980s. Yet, there is the more poignant view that epistemologically, development has actually not been on the agenda since post-independent Africa, because what is termed as development is a mere mimicry of the Western idea of development and the invention of Western social sciences (Ake, 1996, 1979). Mkandawire's (2010) force de tour of why Africa needs to run while others walk fits into the latter thinking that what many post-independent African leaders wanted to do was to catch up to the West, as if that was the end goal of a society in terms of socioeconomic transformation. Yet, as Sen (2000) would argue at a global level, development in its proper sense transcends economic growth to creating conditions that allow for freedom of thought and expression, even if such thought is contrary to the view of the ruling elites in power. What has been missing in these strands of scholarly debates is the role of indigenous knowledge in fostering socioeconomic transformation in Africa. Although there have been literature that identified and indeed called for recentering of African indigenous knowledges in fostering socioeconomic development, see (Oloruntoba, Afolayan and Yacoob-Haliso, 2020, Hountondji, 1995), they have not directly explored the possibilities in applying African indigenous knowledge in relation to reconceptualizing development itself, at least from the perspectives of Africans, looking inward and recentering the agency and voices of Africans themselves in designing and implementing policies around socioeconomic development.

Despite the usefulness of these scholarly interventions, they are still ensconced under developmentalism as an end goal of society. Developmentalism itself has been criticized for its extraversion, path dependency, and lack of rootedness in the local realities of the postcolony. In what Adedeji (2002) calls the development merchant system, developmentalism and its dependency on foreign aid limit the agency of Africans in developing their own paths and policies. Worse, most of the foreign aid, which is ostensibly meant for transformation in Africa, ends up being used to pay Western-based consultants. The aid also fosters corruption and stifles initiative for critical thinking and accountable governance (Moyo, 2009).

This chapter departs from the different strands of literature above and analyzes the possibilities, challenges, and potentials of leveraging African indigenous knowledge to foster socioeconomic transformation on the continent. In doing this, we started by analyzing the dominant Eurocentric development paradigm that has pervaded the continent since colonial times through to the period of independence and thereafter. In taking this approach, we provided some data that shows Africa's current socioeconomic conditions in terms of Gross Per Capita income, poverty, inequality, access to education, health, as well as the freedom to voice out anti-government views. In undertaking this task, we also show the inherent contradictions and limitations in the dominant approaches that continue to keep Africa in chains (Ayittey, 2005), not only of debt but of the epistemology of what constitutes development. Apart from showing the contradictions, we show the limitations of extraversion as a knowledge system that continues to shape Africa's path to economic transformation.

In the section that follows, we present a case for the reconceptualization of development in ways that can fit into the local realities and expectations of Africans. This reconceptualization is critical because the Western conception of development has become hegemonic, leaving little or no room for original thinking on how people can live their lives. Development has also become like a mirage in which the closer one is to reaching it, the farther it appears to be. In the next section, we foreground the salience of African indigenous knowledge in its multidimensional uniqueness and application in the search for the transformation of society. In this section, we also use the history of African entrepreneurship as a starting point for analyzing the possibilities of leveraging indigenous knowledge both at the formal and non-formal levels to address the challenges of poverty and joblessness, among others. The various sections in this chapter are theoretically grounded in decolonial and Afrocentric theoretical standpoints that recenter Africa's scholarship, voices, and experiences. In other words, we discuss the salience of African indigenous knowledge as a valid knowledge system which possess in it resources, ideas, and perspectives that can foster socioeconomic transformation beyond what we have come to accept of what development connotes and how it can be achieved. By exploring the possibilities inherent in Africa's indigenous knowledge on innovation, technology, artisanal mining, and other sectors, this chapter complements other chapters in this book but departs from what we have accepted as development in Africa.

Development and Its Illusions in Africa

The colonial project of modernization and civilization in Africa was abruptly aborted after the Second World War, with the ascendancy of the United States of America as a world superpower and the rise of nationalist resistance in different parts of Africa. As scholars have shown, despite the official pretension to the contrary, the colonial project was decidedly exploitative, extractive, and disruptive of the political economy of the colonized societies (Ake, 1981; Rodney, 2018). As Acemoleglu, Johnson, and Robinson (2001) argue in their controversial Nobel Prize-winning work on institutions, in societies where colonial governments established extractive institutions, only a few benefits, and these societies remain underdeveloped today. Although this view has been criticized for reifying Western institutions over non-Western institutions, their analysis correctly shows that colonial institutions, especially in much of Africa, were deliberately extractive and created the conditions that persist in contemporary Africa.

With Ghana's independence from the British in 1957, the door was widely opened for other countries to gain their independence. Prominent leaders among the newly independent countries, such as Kwame Nkrumah, Kenneth Kaunda, Julius Nyerere, and Uhuru Kenyatta, among others, developed political and economic thoughts through which they hoped to transform their societies. Towards achieving socioeconomic transformation, they invested heavily in social policies such as massive free and compulsory education at the primary and high school levels, primary health care, and pension (see Adesina, 2023; Oloruntoba, 2023; Aina, 2004; Afolayan and Ojebode, 2023). In Nkrumah's Ghana, the government adopted African socialism as an ideological blueprint that undergirded the active involvement of the state in the establishment of industries, major infrastructural projects such as dams and roads, as well as state-owned financial institutions to support an emergent African merchant class. Under Julius Nyerere in Tanzania, Ujamaa was developed as an egalitarian economic blueprint to foster equal and even development by returning the population to village life and communal work under the villagization project, promoting agriculture, creating access to education and health, and ensuring state-led industrialization. African leaders also adopted Import Substitution Industrialization strategies to promote industrialization by reducing dependence on imports (see Nason'go, 2018). Figure 10.1 shows the growth in Gross Domestic Product as well as Gross Per Capita Income from the 1960s. It also shows a rapid increase in GDP growth in 1960, followed by a drastic decline in 1964 and another growth from 1966 to 1968. This pattern of intermittent growth and fall in GDP has continued till contemporary times. What is obvious is that the growth pattern is neither high enough nor sustained long enough to generate any meaningful impact. The growth has also largely been fueled by the extraction of minerals and export of primary products, although services have made some contributions. Apart from a lack of value addition, which could have bolstered competitiveness and created multiplier effects in the rest of the economy, exports of primary products and unprocessed minerals inevitably make African economies vulnerable to the fluctuations in international markets.

The lack of capacity to mobilize capital inevitably led African countries to huge debt burdens from the 1980s. By this time, the world has fallen under the neoliberal economic doctrine, which sees the market as the best agent for allocating economic resources (Harvey,2007). As a way out of the economic challenges, several African countries sought and received loans from both public and private financial markets from the West, such as the London and Paris Clubs, as well as from multilateral lending institutions such

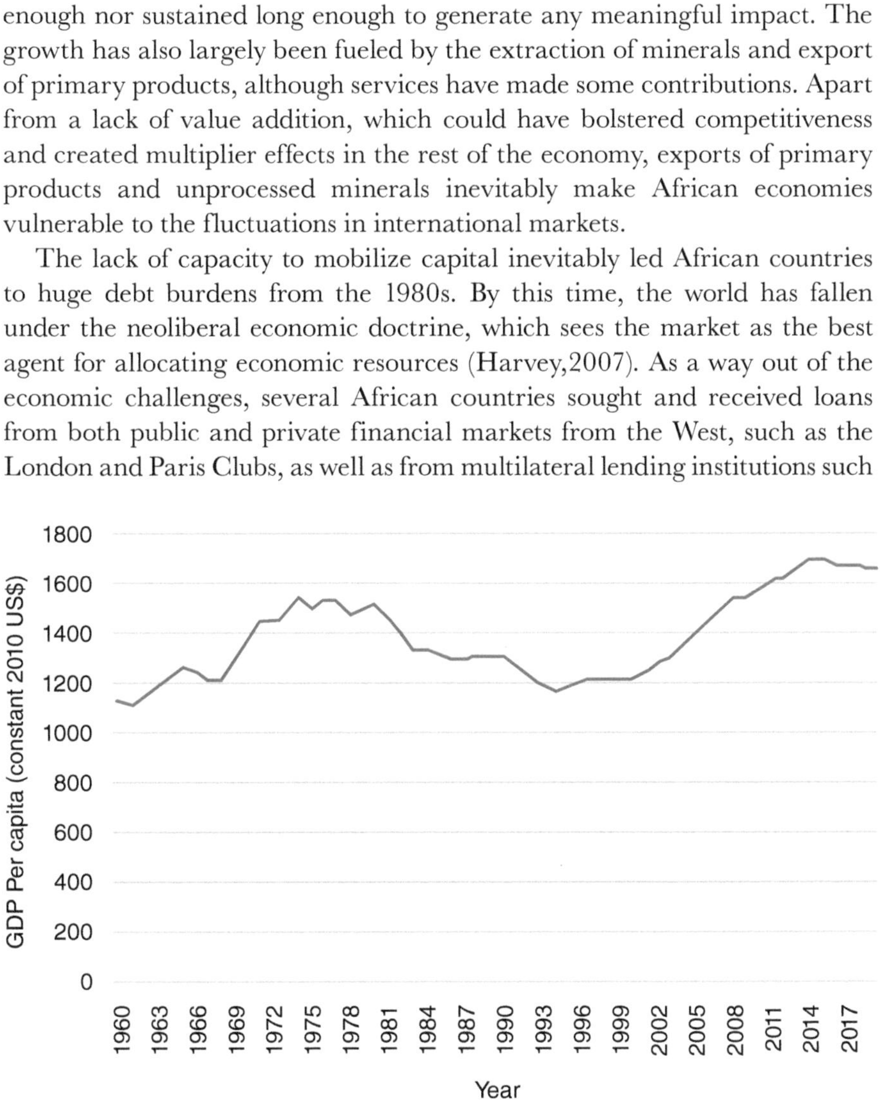

Figure 10.1 The growth in Gross Domestic Product as well as Gross Per Capita Income from the 1960s

as the IMF and the World Bank. To qualify for the loans, several conditions were imposed. In what Williamson (1990) calls the Washington Consensus, a program of structural adjustment was introduced as a package of reforms for the continent. The package of reforms included privatization of state-owned enterprises, many of which have been mired in corruption, deregulation of key sectors of the economy to allow the invisible hands of demand and supply, liberalization of trade and investment regimes, among others (Olukoshi, 1998; Soludo and Mkandawire, 1999). Because these loans came at very high prices and due to the corruption in the management of the loans, there was no evidence that the various governments used them to achieve the purposes for which they were borrowed.

From the late 1980s, the high debt burdens and the cost of servicing them started to affect government deliverables in social services. This led to campaigns for debt relief and debt forgiveness by political, academic, entertainment personalities, and non-government organizations such as Tony Blair, former Prime Minister of Britain, Jeffrey Sachs, and Bruno, and Jubilee 2000, among others. In response to these demands, the IMF introduced the Highly Indebted Poor Countries Debt Relief programs in which billions of dollars were either written off or restructured (IMF 2005). A set of other market-oriented programs, such as Poverty Reduction Strategies for Poor countries were introduced by the IMF to help African countries restructure their economies. Across the continent, various market-oriented programs such as Growth, Employment and Retribution in South Africa, National Economic Empowerment and Development Strategy in Nigeria, among others, were introduced to bolster growth. To some extent, these programs were able to stimulate some level of growth, but because they lacked structural transformation and rootedness in the indigenous knowledge of the various countries, the growth was short-lived. Nor were they able to lead to any significant shift in unemployment and poverty levels. What has emerged is that Africa has entered new levels of debt accumulation, both with previous and new lenders. In what Oloruntoba (2021) refers to as going back to the well, most African countries are now in the throes of a new debt burden to both old and new lenders, especially China. As the figures show (Figures 10.2–10.4), neither the previous growth trajectories of the first decades nor the ones of the early 2000s were able to shift the poverty level in a significant way.

The methodological approach for computing growth rate in Africa has been based on the economic output of the formal sector. Yet in many countries in Africa, the so-called informal sector is the highest employer of labor. Beyond that, this sector provides access to means of livelihood for women as well as contributes to the ability of parents to send their children to school, both locally and abroad. Lack of inclusion of the output from the

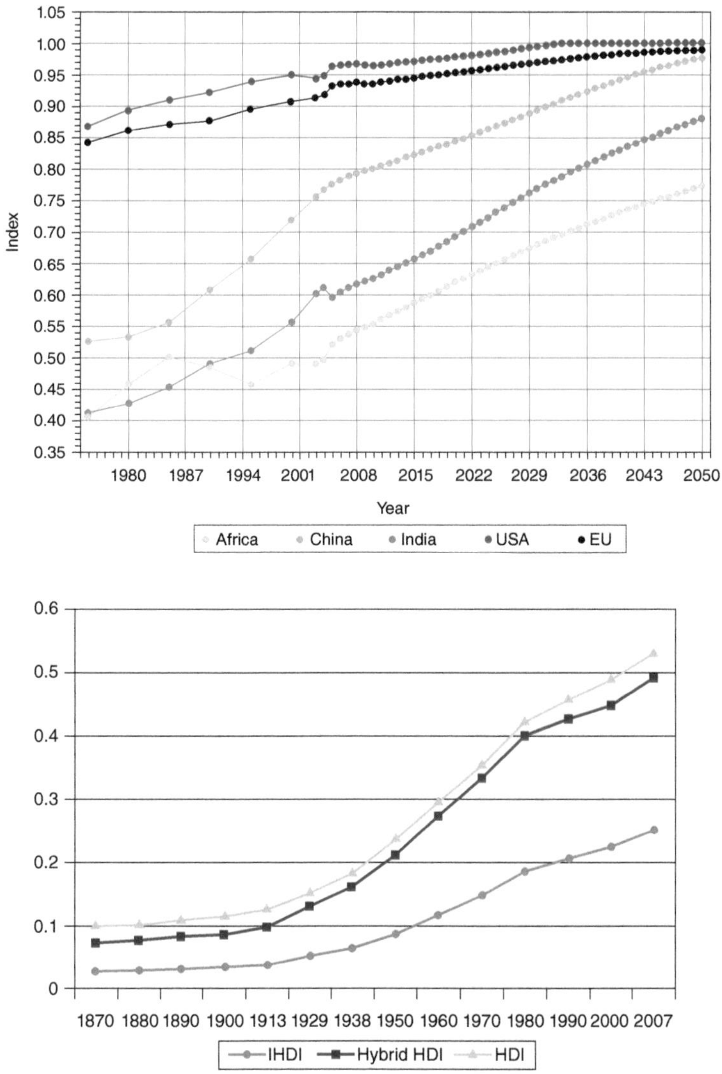

Figure 10.2 (a,b) A comparison of the human development index between Africa, China, India, the United States of America, and the European Union. It shows that whereas Africa and India were at relatively the same Human Development Index at 0.40 before 1980, India has comfortably leaped forward to 0.70 by 2022 and is expected to reach 0.50 by 2050, while Africa was at 0.62 by 2022 and is expected to reach 0.75 by 2050. While the US and EU have prior advantages of serving as the core of global capitalism, with innovation and technology to match, China and India are relatively late industrial societies that adopted a mix of local and external knowledge systems to achieve growth records that not only show in GDP but also in human indices such as improvement in health, education, access to water, and overall reduction in poverty. Source: Human Development Index (HDI), Africa in a global context. Created by Jakkie Cilliers, Barry B. Hughes, and Jonathan Moyer.

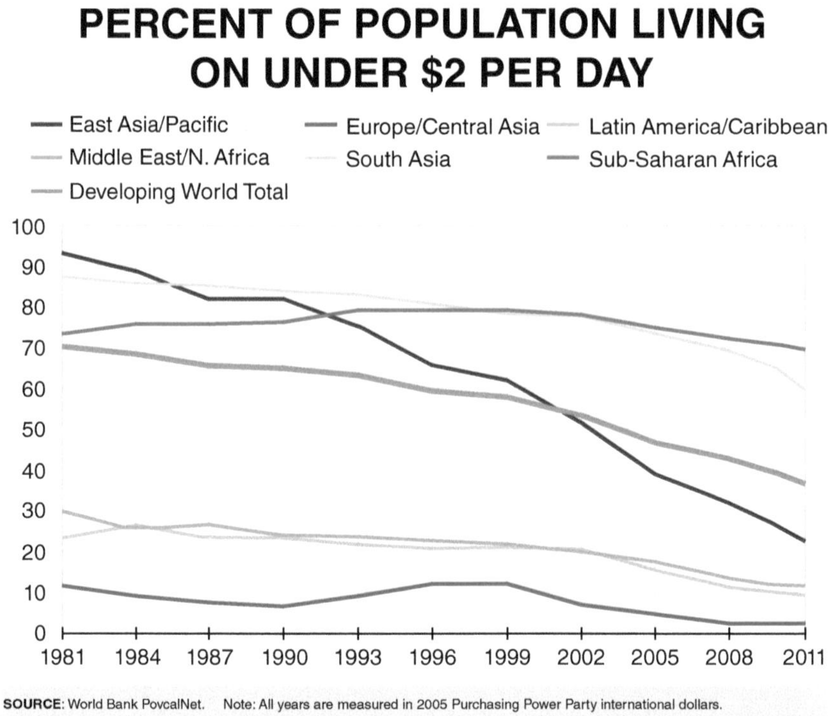

Figure 10.3 Although the poverty rate in Africa has gone down, the number of African people living in poverty has increased. The World Bank attributes this contradiction to the growth in population in Africa. This argument can only be partially true because both India and China have higher populations than Africa until at least two years ago, but they have managed to lift millions out of poverty due to the mix of home-grown and orthodox approaches that they have adopted. Source: World Bank PovcalNet, 2015.

Informal sector has implications for getting accurate data on the rate of poverty in Africa. Thus, there are recent concerns among scholars that the mischaracterization and labelling of Africa's economy under the binary of formal and informal economies constitutes a great challenge, which should be part of the ongoing efforts at decolonization and recentering of Indigenous knowledge. As Akinboye and Oloruntoba (2025) argue, given the historical importance and cultural connections between Africans' ways of life and their economic activities, as well as the roles that these activities continue to play in ensuring sustainable livelihood, it is imperative to rethink the nomenclature of informality. The need to change this nomenclature is necessary because the categorization of roadside and street trading, cross-border trading, repair of cars and electronics as informal has resulted in abuse by state officials

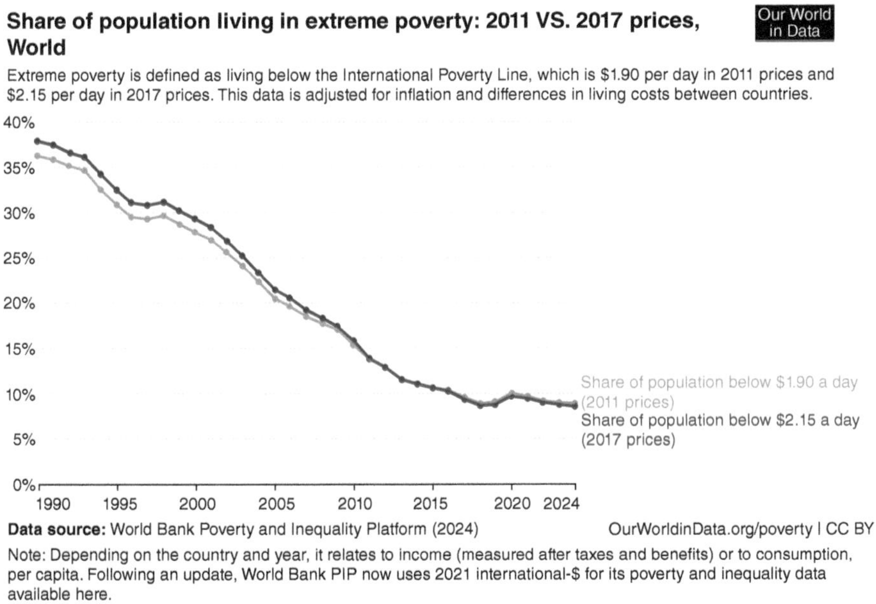

Figure 10.4 The number of poor people who have fallen below the $2.15 poverty line from 1991 to 2019. The figure shows that Asia and Africa have seen the highest number of people falling below the poverty line. However, from 2015, while Asia recorded a smaller number of people, Africa has continued to see a rise in the number of people falling below the poverty line. This raises a lot of questions with respect to data sources, methodology, and the scope of measurement of economic activities. One of the problems of measuring data in Africa is not only the reliability of the data, but also the composition. As Jerven (2013) argues in his celebrated book, no statistical techniques can substitute for partial and unreliable data. Source: Our World in Data.

through extortion, over taxation, harassment on the streets, and neglect by the state. The next section problematizes development and foregrounds the need for indigenous knowledge to achieve sustainable livelihood in Africa.

Who Owns the Development Discourse and Practice in Africa?

One of the most common buzzwords among the ruling elites, in developed and emerging economies, and international institutions is development. The various policies and programs mentioned in the preceding sections were geared towards economic development. Economic development has been conceived as the quantitative and qualitative improvement in the living standards of the citizens (Todaro and Smith, 2010). It also describes a process of improving the economic and social well-being of people in a specific area.

In addition to qualitative and quantitative growth, it includes social and environmental aspects (Ručinská, Müller, Nauerth, 2016). The preoccupation with achieving development and catching up with the West (Mkandawire, 2010) has made African countries pursue an idea that they neither conceived nor own, thus relegating and denigrating their indigenous knowledge, which has been in use for thousands of years. This futile search for development has created conditions of dependency, mimicry, and perpetual frustration and disillusionment. Ake (1996) provokes a deep thinking about the failure of the development enterprise in Africa. He notes that despite three decades of preoccupation with development in Africa, the economies of most African nations are still stagnating or regressing. For most Africans, incomes are lower than they were two decades ago, health prospects are poorer, malnourishment is widespread, and infrastructures and social institutions are breaking down. This observation led him to argue that the problem is not that development has failed, but that it was never really on the agenda.

It has been almost three decades since Ake made these observations. Yet the conditions remain the same today. Poverty remains rife, despite growth in GDP, while the rate of unemployment, especially among the youth, remains stubbornly high despite growth in the services sector. The contradictions of touting development as the main aim for African governments under a global capitalist and a knowledge system of which they were neither the main nor principal actors can be interpreted at various levels. The first is at the level of conceptualization. As we mentioned above, African countries transitioned from a colonial system of extraction to a neocolonial system of seeking to resemble the core of capitalism, where their labor and resources have been directed during the colonial period (see Nkrumah, 1966). By not looking inward to reengage with the precolonial knowledge and economic systems that the colonialists violently disrupted and recreated to serve their own interests, African leaders lost an opportunity to use available resources to transform their societies. In this process, the vast opportunities available in indigenous agricultural practices such as crop rotation, indigenous health practices, traditional governance systems based on the decentralization of power, partnership with indigenous communities, and refocusing social and economic policies from urban to rural areas were lost. Although there have been instances where the postcolonial state tried to focus on rural areas, these did not go far enough, nor did they receive enough financial support to achieve their goals. A clear example of this was the Ujamaa under Julius Nyerere, the former president of Tanzania. Despite or because of the merit of this program and its rootedness in African indigenous knowledge of communal living, focus on rural areas, the World Bank failed to provide the needed capital because of its alleged socialist orientation (Payer, 1983).

Of course, scholars have identified other factors that led to the failure of the program. These include exploitation and domination of the poor peasantry by the bureaucrats and the rich capitalist farmers; the inadequate endowment of the poor peasantry in skills and education, the myth according to which communal traditions favored socialism, and the great food shortage which hit the country between I973 and 1975 (Ergas, 1980). However, as Payer (1983:388) argues, 'there is general agreement that the World Bank was never enthusiastic about the idea of communal production, 'which was a critical aspect of the ujamaa program. This lack of interest and the accompanying difficulty in accessing funds for homegrown programs are one of the challenges faced by post-independent African countries.

Closely related to the above is the issue of ownership of the various development policies. Harper-Shipman (2019) provides an insightful analysis of how donor countries have mastered the language of ownership in a manner that means something different when it comes to African agency and space to own the nationalist programs that they had hoped to transform the societies. In his review of this book, Gordon (2020:621) argues

> that the ownership paradigm co-opts the language of partnership, country ownership, local knowledge, and stakeholder participation to reanimate the spirit of competition, private profit, measurable results, hierarchies of knowledge, and co-optation. At work in the transformation is the further disemboweling of the linguistic coherence of the political aspiration of African national self-determination.

What emerged from the above is that though Africans were made to believe they own the development agenda, the conceptualization, designs, and the end goals of such an agenda were not their own. Thus, while the state elites and even civil society organizations might have been consulted, such as in the processes leading to the design of the Poverty Reduction Strategy Papers, the templates, metrics, and conditions were developed by the IMF.

The third challenge is the lack of sufficient engagement with the traditional institutions at the local level. Whereas the various traditional institutions played very important roles in the precolonial period, and the long period of colonialism did irreparable damage that continues to define how the state and the people relate with these institutions, which have undermined their relevance in contemporary times (Mamdani, 1996; Megisteab, 2019). The disengagement from the traditional economic and political institutions corresponds with the loss of the indigenous knowledge systems in these localities. Thus, rather than co-creating knowledge with the traditional institutions, knowledge holders in the communities, the political elites in

Africa have consistently worked with the World Bank and IMF staff and consultants to design programs that they consider appropriate for the people. In his critique of these institutions, Stiglitz (2002) shows how IMF officials usually meet with Central Bank officials from developing countries in five-star hotels to review the macroeconomic performance of these countries, and make recommendations for reform, without doing so much as to directly engage with the supposed beneficiaries of such interventions. This is especially true for people in rural areas, who are often neglected by the state. The extraversion of the knowledge and approaches to socioeconomic development in Africa requires epistemic decolonization and the recentering of indigenous knowledge in Africa. The next section addresses this issue using the economic history of Africa and African entrepreneurship as points of entry.

Decolonizing Socioeconomic Development Through African Entrepreneurship

The past and ongoing debates on decolonization in Africa have been focused on the search for epistemic justice and freedom (Ndlovu-Gathseni, 2018, 2015; Oloruntoba, 2014; Ake, 1979). The link between the epistemic disruption caused by colonialism and the continuing challenges of poverty has not been addressed. In this section, we argue that the term socioeconomic development needs to be reconceptualized to reflect the knowledge, needs, and interests of Africans. As discussed above, development has often been conflated with economic growth, yet as Seers (1972/2012) argues, there is a difference between economic growth and economic development. For him, rather than just measuring growth based on economic output, the concern should be about people. Seers prioritize people and their capacity to realize human personality as the ultimate goal of development in all countries. Thus, to be complete, development should be about the realization of human dignity. This dignity encompasses self-consciousness, pride in one's identity, culture, history, and language. To be sure, Kwame Nkrumah, former Ghanaian President, built on the earlier advocacy for Black pride and African Personality developed by Pan-Africanists like Edward Blyden, Casely Hayford, George Padmore, and Nnamdi Azikiwe to promote the philosophy of African Personality in post-independent Ghana in 1957 (Amoh, 2022). Nkrumah defines African Personality as a set of humanist values rooted in precolonial African societies (Nkrumah, 1970). According to Amoh (2022), these humanist principles embodied the institutions and mores of Africans that sought to improve their lives economically, politically, and culturally. As Amoh further notes, 'the pursuit of a new African Personality was essential to Pan-African goals of African liberation, unity, and cultural pride, among others' (p.35).

The African Personality foregrounds what development should be about, both in cultural expression and connection among Africans. By being connected with Pan-Africanism, it transcended the strictures of colonial boundaries upon which post-independent African leaders have continued to seek transformation of their societies. To cite Amoh again, the 'African Personality framed postwar Pan-Africanism and African nationalists' efforts to re-imagine Africa and Africanness in postcolonial Africa' (p.35). The re-imagination of Africa is crucial because colonialism redefined Africa in the image of the colonialists for the purpose of serving the economic needs of the metropoles in London, Paris, Berlin, Portugal, and Brussels.

Despite the criticism levelled against the concept of African Personality, such as being a mere slogan, archaic, or being too essentialist while ignoring what Mazrui (1986) calls the triple heritage of indigenous, Islamic, and Western heritage of Africa, it was an original and intentional idea geared towards decolonization and could have contributed to reimagining better approaches to achieving socioeconomic transformation in Africa. Unfortunately, both domestic and external contradictions led to the military overthrow of Kwame Nkrumah in 1966. The sudden end of his government created a huge vacuum in the search for identity, cultural pride, and adoption of a distinctly African approach to achieving socioeconomic transformation that has hardly been filled in Africa today. The reinvention of this idea by President Thabo Mbeki of South Africa and his associates, like President Olusegun Obasanjo of Nigeria and Abdoulaye Wade of Senegal, under what they called the African Solution to African Problems, did not go far enough because of both domestic and external challenges.

In the context of the theme of this book, decolonizing socioeconomic development would require reimagination of the history, indigenous knowledge system, economy, politics, culture, and identity. With a focus on indigenous knowledge and entrepreneurship, the past can guide the present and the future of socioeconomic transformation. Reimagining Africa in these domains would assist us to know what has been achieved in the past, what is ongoing, and what can be done differently. The general economic history of Africa has been framed around a continent that was essentially rudimentary and limited in participation in global trade and application of technology. Some even argued that because the colonialists did not make Africa's economy in the same structural form as European economies in the 19th century, there were no capitalists on the continent, nor do Africans have the capacity to build businesses. This ahistorical generalization has been faulted. As Ochonu (2018) magisterial edited book on African entrepreneurship clearly shows, the denial of Africa's engagement in entrepreneurship was a mere continuation of Eurocentric negation of knowledge production in Africa.

Ochonu and other contributors in this book show the various dimensions of African's engagement with entrepreneurship from precolonial through colonial to contemporary times. While not dismissing the contributions of non-African scholars to the study of entrepreneurship in Africa, Ochonu departs from the conventional wisdom that entrepreneurship can only be understood from Joseph Schumpeter's or neoclassical economic theories, which conceive entrepreneurship from a purely profit-making perspective. He argues that at the theoretical level, entrepreneurs in Africa make a profit but go beyond profit-making to include care for the community. This point is critical because concerns for the community and the environment are one of the core tenets of indigenous entrepreneurship (Colborne, Peredo, and Henriques 2023). This is underpinned by indigenous knowledge and the philosophy of communal care.

Beyond corporate social responsibility, which companies do to show that they care for their host communities, African entrepreneurs have been historically immersed in their communities through apprenticeship, mentoring young and upcoming entrepreneurs, providing capital for fellow or upcoming entrepreneurs, resolving conflicts, and building networks of trade. Decolonizing entrepreneurship would require a new engagement with the underlying indigenous knowledge basis of entrepreneurship in Africa. While centers of entrepreneurship continue to spring up across universities in Africa, the majority of them are based on the Eurocentric idea of entrepreneurship, which reifies profit-making, self-fulfillment at the expense of the communal benefit. It is also ironic that, except for the Indigenous Knowledge System Center at the Northwest University Mafikeng, South Africa, where there has been a deliberate effort to apply indigenous knowledge to entrepreneurship initiatives, the few indigenous knowledge centers that exist in Africa are focused on the theory of indigenous entrepreneurship. Yet to address the problems of high unemployment in Africa, it is imperative to apply the various indigenous knowledge in establishing entrepreneurial activities.

Given the rich resources in Africa's indigenous knowledge, there are massive opportunities to leverage the wide tapestry of these knowledge systems to foster entrepreneurship. From mining to textiles, craft to agriculture, indigenous knowledge can be applied to generate jobs for men and women across the continent. Following on the heels of the various trade networks such as the Wangari trade networks in West Africa (Ochonu, 2018), trade routes connecting Khoisan, Batswana, Khosa, and Nguni speakers, Eastern and coastal trade networks, as well as long distance trade networks across the Sahara, which facilitated exchange of various goods and commodities such as salt, gold, iron, ivory, beads and other valuable items in precolonial times, contemporary African leaders can unlock massive indigenous entrepreneurship by removing barriers to trade and movement of people.

Studies have shown that Africans want to be free to move freely and trade with one another, but the colonial boundaries created at the Berlin conference and upheld by the postcolonial leaders continue to hinder such human and material exchanges (Gumede, Oloruntoba, and Kamga, 2019). The recent launch of the African Continental Free Trade Area and other various regional integration programs are steps in the right direction. However, these initiatives need more political commitment and citizen buy-in to make them successful (Oloruntoba, 2023). The challenges associated with lack of complementarity in the economies of African countries, lack of adequate infrastructure to connect the continent in meaningful ways, the fear that bigger countries would derive more benefits from the integration arrangements and worse the recurring cycle of xenophobic attacks against other Africans in South Africa, and resulting retaliation against South African companies in Nigeria, Zambia and elsewhere constitute key impediments to using the African economic system to foster indigenous entrepreneurship among the citizens.

Additionally, sustainable economic development can be achieved by revisiting the grammar of informality. As previously mentioned in this chapter, the misuse of the grammar of informality to describe economic systems and activities that are peculiar and consistent with the African indigenous knowledge system has fostered abuses, neglect, and extortion by state officials. The coloniality of language that has defined economic activities in Africa has also affected how the economy itself is organized. In the agriculture sector, for instance, while the massive arable land in Africa is supposed to ensure food sufficiency and security, African leaders who are under the influence of coloniality would rather welcome agribusiness organizations to cultivate farms for exports, rather than produce for local consumption. Ironically, Africa continue to import food from other regions of the world (Amusan and Ajibola, 2019). Worse still, some of the seedlings and methods of farming that have been introduced to Africa have altered the indigenous ways of farming and crop production. The introduction of Genetic Modified (GMO) foods, while raising food productivity for exports, has implications for the health of the population as well as the ecosystems (Tandon, 2015). While it may be expedient to commercialize farming to foreign exchange, ensuring sustainable development and prioritizing the socioeconomic well-beings of Africans, both in urban and rural areas require a new engagement with indigenous knowledge in Africa in ways that would allow for sustainable farming.

Indigenous entrepreneurship opportunities exist in African medicine that have been used and still being used for sustainable livelihoods across Africa. From the native doctors or *sangomas* who consult and divine for people, herbalists who prepare herbs to treat diseases, to traditional midwives who take delivery for pregnant women in ways that align with their spiritual beliefs (Shope, Rabie, Du Perez, and Bester, 2020) Africans have continued to engage with indigenous medicine for sustainable livelihoods even in postcolonial times. Although the commercial activities of indigenous medical practitioners have long been consigned to rural and semi-urban areas, the digital space has changed this very dramatically. Practitioners now use social media to promote their products. The huge African diaspora population has also provided new marketing opportunities for the practitioners to sell their herbs (many of which are well prepared. Besides, Africans are now making efforts to standardize their indigenous medicine by subjecting them to scientific evidence. A successful example of this company is Phytotech, an Accra, Ghana based indigenous led pharmaceutical company that is using herbal evidence- based herbal solutions to cure various diseases such as prostate cancer, foster blood circulation, among others (www.phytotecgh.com). There are several other companies that undertake similar undertakings across the continent. The government can play more roles by providing support in the form of low interest loans to these types of companies to expand their operations, train younger generations of Africans who would be engaged in these lines of jobs. This will not only help to preserve the indigenous knowledge but generate value for all stakeholders.

Additionally, indigenous knowledge can be leveraged to foster climate change adaptation through the various practices such as forest conservation, groves, and protected areas. Indigenous knowledge can also foster climate change through water sector adaptation (Filho, et al, 2022). In their study of how water adaptation is used to manage climate change, Zvobgo, et al (2022) found that irrigation, rainwater harvesting, water conservation, and ecosystem-based measures, mainly agroforestry, were the most implemented measures across Africa. These practices, if amplified and adequately harnessed can help to mitigate the challenges of climate change and enhance sustainable livelihood in Africa. Given the cyclical weather conditions and the increasing unpredictability of weather due to climate change, rainwater harvesting, water conservation can help farmers to cultivate their crops all the year round.

Conclusion

This chapter foregrounds indigenous knowledge as a way of setting a new agenda for socioeconomic and sustainable development in Africa. We started by problematizing the word development and locating it within an industry that Adedeji (2002) calls the development merchant system. We also argued that economic growth is not tantamount to economic development. With data and case studies, we showed how poverty continues to define the landscape in Africa despite incidences of economic growth. We also argued that the language of informality has limited the space in which indigenous knowledge can be harnessed to foster transformation. By mislabeling economic activities that provide livelihood for more than half of the population, African leaders have allowed coloniality to fester for too long. Thus, it is imperative to decolonize economic thinking, language, policies, and practices.

We contend that the neglect of the rich and massive indigenous knowledge resources in Africa, especially in areas of entrepreneurship, is one of the main reasons for the increasing poverty and general underdevelopment in Africa. We showed that indigenous entrepreneurship is not alien to Africa, as the economic history of Africa showed various forms of entrepreneurship activities before the colonial period (Zeleza, 1993; Ochonu, 2018). Indigenous entrepreneurship is different from conventional entrepreneurship because it is not only focused on profit-making but on community interests and preservation of the environment, as well as consideration for the future. Furthermore, we argued that by decolonizing development, Africa can use its massive indigenous knowledge to unleash opportunities in agriculture, traditional medicine, arts and culture, as well as ecotourism, among others. The chapter also showed areas in which indigenous knowledge is being used to foster livelihood, including traditional medicine, traditional midwives, sangomas, and so on. The importance of indigenous knowledge in climate adaptation was also highlighted. As the decolonization debates continue, the governments of African countries would do well by promoting indigenous knowledge through its inclusion in the curriculum from elementary to university levels.

To summarize, this book foregrounds indigenous knowledge as an epistemic resistance against coloniality. Using decolonial and feminist theoretical frameworks, it argues that development must be reconceptualized and interpreted from the perspectives of African indigenous knowledge systems. By amplifying indigenous knowledge, African countries can tap into human and material resources in several sectors such as agriculture, craft, culture, creative health, climate, and tourism to improve livelihood. Rather than continuing to depend on Eurocentric ideas on entrepreneurship, which have been proven to be limited in their capacity to generate jobs and foster inclusive transformation, African societies can look back to the diverse economic system and practices of the precolonial times and adapt them to the modern practices. The revolution in technology and global mobile communication systems has provided massive opportunities to promote indigenous knowledge and enhance sustainable livelihood in Africa. Taking advantage of the complementarity that this combination of knowledge provides requires decolonization of the minds of government officials, the majority of the citizens who are ensconced in coloniality, an intentional shift in the language of informality, infusion of African indigenous knowledges in the curricula, and establishment of centers of African indigenous entrepreneurship across urban and rural areas.

References

Adebanwi, W. (2015). *Authority stealing: Anti-corruption war and democratic politics in post-military Nigeria*. Carolina Academic Press.

Adebanwi, W., & Obadare, E. (2010). Introducing Nigeria at fifty: The nation in narration. *Journal of Contemporary African Studies, 28*(4), 379–405. https://doi.org/10.1080/02589001.2010.512737

Adedeji, A. (2002, April 26–29). *From the Lagos Plan of Action to the New Partnership for Africa's Development and from the Final Act of Lagos to the Constitutive Act: Whither Africa?* [Keynote address]. African Forum for Envisioning Africa, Nairobi, Kenya.

Adesina, J. (2023). Transformative social policy in African context. In S. O. Oloruntoba, F. Bammeke, & O. Durowade (Eds.), *Social policy and transformation in Africa: A festschrift for Professor Omotade Akin Aina* (pp. 57–89). Noirledge Publishing.

Aina, T. A. (2004). Introduction: How do we understand globalisation and social policy in Africa? In T. A. Aina, S. L. Chachage, & E. A. Yao (Eds.), *Globalisation and social policy in Africa*. CODESRIA.

Ake, C. (1979). *Social science as imperialism: A theory of political development*. Ibadan University Press.

Ake, C. (1981). *The political economy of Africa*. Longman.

Ake, C. (1996). *Democracy and development in Africa*. Brookings Institution.

Akinboye, S., & Oloruntoba, S. O. (2025, June 3–6). *Reframing the grammar of economics: Indigenous knowledge and economic transformation in Africa* [Conference paper]. Canadian Association of African Studies Conference, University of Toronto, Toronto, Canada.

Amin, S. (1990). *Maldevelopment: Anatomy of a global failure*. Pambazuka Press.

Amoh, E. (2022). Revisiting Kwame Nkrumah's African personality: The diaspora context and the making of Ghana television. *Ghana Studies*, 25, 33–56.

Amusan, L., & Ajibola, K. S. (2019). Human resources and food insecurity in Africa: Adventure in Africa solution for African problems. *Journal of Gender, Information and Development in Africa, 8*(Special Issue), 65–84. https://doi.org/10.31920/2050-4284/2019/s1n1a4

Ayobami, O., & Afolayan, A. (2023). The sociological imagination, social imagery and the social policy imperative in Nigeria. In S. O. Oloruntoba, F. Bammeke, & O. Durowade (Eds.), *Social policy and transformation in Africa: A festschrift for Professor Omotade Akin Aina* (pp. 115–134). Noirledge Publishing.

Bayart, J.-F. (2009). *The state in Africa: The politics of the belly* (2nd ed.). Polity Press.

Colbourne, R., Peredo, A. M., & Henriques, I. (2023). Indigenous entrepreneurship? Setting the record straight. *Business History, 66*(2), 455–477. https://doi.org/10.1080/00076791.2023.2166034

Collier, P. (2007). *The bottom billion: Why the poorest countries are failing and what can be done about it.* Oxford University Press.
Easterly, W. (2007). *The white man's burden: Why the West's efforts to aid the rest have done so much ill and so little good.* Penguin Random House.
The Economist. (2011, December 3). *Africa rising.*
Ergas, Z. (1980). Why did the Ujamaa village policy fail? Towards a global analysis. *The Journal of Modern African Studies, 18*(3), 387–410.
Fioramonti, L. (2013). *Gross domestic problem: The politics behind the world's most powerful number.* Zed Books.
Gordon, J. A. (2020). Rethinking ownership of development in Africa: An introduction. *Journal of Asian and African Studies, 55*(5), 621–624.
Gumede, V., Oloruntoba, S. O., & Kamga, S. D. (2019). *Regional integration and migration in Africa: Lessons from Southern and West Africa.* Brill in association with CODESRIA.
Harper-Shipman, T. D. (2019). *Rethinking ownership of development in Africa.* Routledge.
Harvey, D. (2007). *A brief history of neoliberalism.* Oxford University Press.
Hountondji, P. J. (1995). Producing knowledge in Africa today: The second Bashorun M. K. O. Abiola distinguished lecture. *African Studies Review, 38*(3), 1–10. https://doi.org/10.2307/524790
International Monetary Fund. (2011). *World economic outlook.* International Monetary Fund.
Jerven, M. (2013). *Poor numbers: How we are misled by African development statistics and what to do about it.* Cornell University Press.
Joseph, R. A. (2014). *Democracy and prebendal politics in Nigeria: The rise and fall of the Second Republic.* Cambridge University Press.
Leal Filho, W., Barbir, J., Gwenzi, J., Ayal, D., Simpson, N. P., Adeleke, L., Tilahun, B., Chirisa, I., Gbedemah, S. F., Nzengya, D. M., Sharifi, A., Theodory, T., & Yaffa, S. (2022). The role of Indigenous knowledge in climate change adaptation in Africa. *Environmental Science & Policy*, 136, 250–260. https://doi.org/10.1016/j.envsci.2022.06.004
Mamdani, M. (2018). *Citizen and subject: Contemporary Africa and the legacy of late colonialism.* Princeton University Press.
Mazrui, A. A. (1986). *The Africans: A triple heritage.* BBC Publications.
Mengisteab, K. (2019). Traditional institutions of governance in Africa. In *Oxford Research Encyclopedia of Politics.* Oxford University Press. https://doi.org/10.1093/acrefore/9780190228637.013.1347
Mignolo, W. D. (2009). Epistemic disobedience, independent thought and decolonial freedom. *Theory, Culture & Society, 26*(7–8), 159–181.
Mkandawire, T. (2010). *Running while others walk: Knowledge and the challenge of Africa's development.* London School of Economics and Political Science.
Mkandawire, T., & Soludo, C. C. (1999). *Our continent, our future: African perspectives on structural adjustment programs.* CODESRIA, Africa World Press, & IDRC.
Moyo, D. (2009). *Dead aid: Why aid is not working and how there is another way for Africa.* Douglas & McIntyre.
Ndlovu-Gatsheni, S. J. (2013). *Coloniality of power in postcolonial Africa: Myths of decolonization.* CODESRIA.
Nkrumah, K. (1970). *Consciencism: Philosophy and ideology for decolonization.* Monthly Review Press.

Ochonu, M. (Ed.). (2018). *Entrepreneurship in Africa: A historical approach*. Indiana University Press.

Oloruntoba, S. O. (2014). Social sciences as dependency: State apathy and the crisis of knowledge production in Nigerian universities. *Social Dynamics, 40*(2), 338–352. https://doi.org/10.1080/02533952.2014.946268

Oloruntoba, S. O. (2023a). The political economy of the African Continental Free Trade Area and structural transformation in Africa. *Journal of Contemporary African Studies, 41*(4), 360–376. https://doi.org/10.1080/02589001.2023.2214715

Oloruntoba, S. O. (2023b). The political economy of social policy and transformation in Africa. In S. O. Oloruntoba, F. Bammeke, & O. Durowade (Eds.), *Social policy and transformation in Africa: A festschrift for Professor Omotade Akin Aina* (pp. 90–114). Noirledge Publishing.

Oloruntoba, S. O., Afolayan, A., & Yacob-Haliso, O. (Eds.). (2020). *Indigenous knowledge systems and development in Africa*. Springer.

Oloruntoba, S. O. (2016). Beyond the fetishism of gross domestic product: (Mis)governance and the challenges of poverty reduction in Nigeria. *Africa Insight, 46*(2), 1–27.

Olukoshi, A. O. (1998). *The elusive Prince of Denmark: Structural adjustment and the crisis of governance in Africa* (NAI Research Report No. 104). Nordic Africa Institute.

Ouko, K., & Odiwuor, M. (2023). Contributors to the looming food crisis in sub-Saharan Africa: Opportunities for policy insight. *Cogent Social Sciences, 9*(1). https://doi.org/10.1080/23311886.2023.2173716

Robinson, W. I. (2004). *A theory of global capitalism: Production, class, and state in a transnational world*. Johns Hopkins University Press.

Rodney, W. (2018). *How Europe underdeveloped Africa*. Verso.

Ručinská, S., Müller, R., & Nauerth, J. A. (2016). Economic development. In A. Marciano & G. Ramello (Eds.), *Encyclopedia of law and economics*. Springer. https://doi.org/10.1007/978-1-4614-7883-6_153-1

Seers, D. (2012). Economic growth: What are we trying to measure? In R. Jolly (Ed.), *Milestones and turning points in development thinking*. Palgrave Macmillan. https://doi.org/10.1057/9781137271631_62

Sen, A. (2000). *Development as freedom*. Anchor Books.

Shopo, H., Rabie, T., Du Preez, A., & Bester, P. (2020). Experiences of women receiving maternal care regarding cultural practices in selected public hospitals in the North West Province, South Africa. *International Journal of Africa Nursing Sciences*, 13, Article 100241.

Stiglitz, J. E. (2002). *Globalization and its discontents*. W. W. Norton.

Taylor, I. (2014). *Africa rising?: BRICS and diversifying dependency*. James Currey.

Tandon, Y. (2015). *Trade is war: The West's war against the world*. OR Books.

Williamson, J. (1990). What Washington means by policy reform. In J. Williamson (Ed.), *Latin American adjustment: How much has happened?* (pp. 7–20). Institute for International Economics.

World Bank. (1981). *Accelerated development in Sub-Saharan Africa: An agenda for action*. World Bank.

World Bank. (2024). *World Bank* report on Africa. World Bank.

Zeleza, P. T. (1993). *A modern economic history of Africa* (Vol. 1). CODESRIA.

Zvobgo, L., Johnston, P., Williams, P. A., Mugandani, R., Nhapi, I., & others. (2022). The role of Indigenous knowledge and local knowledge in water sector adaptation to climate change in Africa: A structured assessment. *Sustainability Science*, 17, 2077–2092.

INDEX